Solar Power for Beginners

Basics, Design and Installation of a Solar Panel System. The Complete Guide for Your Off-Grid Home.

Lawson Lawrence

© Copyright 2020 by Lawson Lawrence. All right reserved. The work contained herein has been produced with the intent to provide relevant knowledge and information on the topic on the topic described in the title for entertainment purposes only. While the author has gone to every extent to furnish up to date and true information, no claims can be made as to its accuracy or validity as the author has made no claims to be an expert on this topic.
Notwithstanding, the reader is asked to do their own research and consult any subject matter experts they deem necessary to ensure the quality and accuracy of the material presented herein.

This statement is legally binding as deemed by the Committee of Publishers Association and the American Bar Association for the territory of the United States. Other jurisdictions may apply their own legal statutes. Any reproduction, transmission or copying of this material contained in this work without the express written consent of the copyright holder shall be deemed as a copyright violation as per the current legislation in force on the date of publishing and subsequent time thereafter. All additional works derived from this material may be claimed by the holder of this copyright.

The data, depictions, events, descriptions and all other information forthwith are considered to be true, fair and accurate unless the work is expressly described as a work of fiction. Regardless of the nature of this work, the Publisher is exempt from any responsibility of actions taken by the reader in conjunction with this work. The Publisher acknowledges that the reader acts of their own accord and releases the author and Publisher of any responsibility for the observance of tips, advice, counsel, strategies and techniques that may be offered in this volume.

Table of Contents

Introduction 1
Chapter 1: Learning About Solar Power 4
    Common Sources of Energy 4
        Oil 6
        Coal.. 9
        Gas..10
        Hydroelectricity..12
        Nuclear Energy...13
        Geothermal Energy 15
        Biomass Energy..16
        Wind Energy...16
    Describing Solar Power 18
        The Benefits of Solar Power 19
        Why is Solar the Superior Renewable Energy Source 20
        Solar Power and Solar Energy 22
        History of Solar Power.. 23
    Innovative Solar Products 25
        Solar Paint 26
        Solar Windows... 26
        Solar Cars 26
        Solar Roads...27

Chapter 2: Setbacks of Solar Power...................... 28
    The Cost of Solar Panels 28
        Review Your Electric Bill... 29
        Evaluate Your Sunlight Exposure............................... 30
        Estimate Residential Solar Panel Cost 30
        Look for Incentives 31
        The Option to Lease 31
    Other Drawback of Solar Power 31
        Location Means A Lot... 32
        Installation Areas 32
        Reliability.. 32

- *Associated Pollution* 33
- Common Complaints About Solar Panels 33
  - *Snail Trails* 33
  - *Birds* 34
  - *Inverter Issues* 34
  - *Faulty Wiring* 35
  - *Roof Issues* 35
- Make Sure Your Roof is Ready 35
  - *Does Your Roof Need Repairs?* 35
  - *What is the Shape of Your Roof?* 36
  - *How Much Weight Can Your Roof Handle?* 36
  - *Where Will the Water Go?* 37

Chapter 3: The Initial Process 38

- Solar Power Equipment 38
  - *Solar Panels* 40
  - *The Solar Inverter* 41
  - *Solar Racking* 44
  - *Solar Performance Monitoring* 44
  - *Solar Storage* 45
- Determining the Number of Solar Panels Needed 46
  - *How Much Solar Power Will You Use?* 47
  - *How Many Watts of Energy Do You Currently Use?* 47
  - *How Much Sunlight Can You Expect?* 48
  - *What Affects Solar Panel Efficiency?* 48
  - *Does Solar Panel Size Play a Role?* 49
- What's With the South? 50
  - *Why is Direction Important?* 50
  - *What About the Angles?* 50
- Protecting Your Equipment 51
  - *Protecting Solar Panels from Hail* 51
  - *Add Solar Panels to Your Insurance:* 53
  - *Damage from Falling Debris* 54
  - *Water Damage to the Panels* 55
- Protection From Thieves 55

Buying Used Solar Panels? 56
   Lower Energy Output *56*
   Loose Connections *57*
   Damaged Panels *57*
   Unsteady Current *57*

Chapter 4: Installation 59

Should You Hire a Pro 59
   Right Method of Installation *59*
   License and Permits *60*
   Manuals and Precautionary Measures *60*
   Electrical Hazards *61*
   Maintenance and Warranty *61*

Getting the Proper Permits 62
   Research Your Local Regulations *62*
   Determine What Documents You Need *62*
   Keep the Government Involved *63*

Installing the System 63
   Step 1: Finding the Best Location *64*
   Step 2: Building the Platform *64*
   Step 3: Mount the Panels *64*
   Step 4: Proper Wiring *65*
   Step 5: Ground the Panels and the Mounting System *66*
   Step 6: Connect the Electrical Components *66*

Troubleshooting and Continued Maintenance 68
   The Wiring is Loose *68*
   The System is Overheating *68*
   The System is Dirty or Damaged *69*

What About in the Ground 70

Chapter 5: The Best Products 73

Finding Who's the Best 73
Some Questions to Ask 74
   How Much Money Will You Save Over 20 Years? *74*
   How Much Will You Pay Upfront? *75*
   Will the System Increase Property Value? *75*

    Questions for the Installer 75
    Questions to Ask About the Warranty and Replacement Procedures 76
    Questions Regarding Liability and Insurance 76

The Best Solar Manufacturers 76
    Auxin Solar 77
    Heliene 77
    Seraphim Solar USA 78
    Solaria 78
    SunPark Technology 78
    Solar World Americas 79
    Panasonic 79
    JinkoSolar 80
    First Solar 80
    LG Solar USA 80
    JS Solar 81
    Canadian Solar 81
    Highest Efficiency Solar Panels 82
    Solar Panels With the Best Temperature Coefficient ... 82
    Material's Warranty 82

Chapter 6: What if You Move? 84

Before Making the Move 84
    Are You Selling Your Property? 85
    Location and Logistics 85
    Rules and Regulations of New Location 85
    Sun Availability 86
    Potential Damage 86

Safe Moving Practices 87
Warranties 90
    Manufacturer's Warranty 90
    Solar Installer Guarantee 91
    A Solar Panel Warranty Becoming Voided 91

Chapter 7: Who Does Solar the Best? 93

Which Countries Do it the Best? 93
    Germany 93

 China *94*
 Japan *94*
 Italy *95*
 United States *95*
Best Solar Power in the United States 96
 California *96*
 Arizona *96*
 North Carolina *97*
 New Jersey *97*
 Nevada *97*
 Massachusetts *97*
 New York *97*
 Hawaii *98*
 Colorado *98*
 Texas *98*
Conclusion **99**

Introduction

Congratulations on purchasing Solar Power for Beginners, and thank you for doing so. You may have heard the phrase solar power on several occasions but not really know what it means or entails. A significant push has been happening to make this a common energy source over other options that are perceived to be more dangerous, expensive, and harmful to the environment. Solar power is considered a superior method in all of these and many more aspects. Despite more people becoming aware of solar power, it is still not accepted by the mainstream. This could be because of misinformation or even a lack of information. Sometimes, people are willfully ignorant or just have not had the information presented to them in an appealing manner.

My goal is to change that with this book. I am passionate about solar power as an energy source and the countless benefits it can provide for people. I truly feel that the more people who learn about the true nature of it, the more accepted it will become by the general public. I also believe that once people get to experience it, they will never want to go back to the older methods.
I'd be glad to hear your thoughts once you've read the book, please let me know by leaving a short review on Amazon!

The following chapters will discuss what solar power is in great detail. We will also cover the history behind it, which will show that it's not a new phenomenon. After reading about solar power, the benefits will become quite apparent. After getting you excited about the concept, I will discuss how you can get started with the right equipment and strategies. You will have to make a significant financial commitment in the beginning, but you will save these abundantly down the line. This book will provide step-by-step guidance to educate you on how to get started and try to clear up any confusion that might. The better you can set up the correct equipment, the more beneficial solar power will be.

Furthermore, I will go over the advantages and disadvantages of different solar products so that you can make the most informed decision possible. There are certain locations in the world, because of the nature of the weather conditions, where solar will be more advantageous than others. However, you are not precluded from the benefits it provides. Solar power is growing in popularity and can become the main energy source throughout the world as positive qualities become more apparent. Several households and companies have already jumped on the bandwagon. I believe that you will understand why after reading this book.

Another great thing is, if you move, you can take your equipment with you. I will cover safe moving practices you can use to transfer your material as safely as possible without doing damage. So, if you are ready to enter the solar power world, then keep reading. I believe the only thing you will be wondering is why you did not make the transition sooner.

Many books are available regarding the topic of solar power and all that it entails. I want to show gratitude to all of you for choosing this one. Every effort was made to make sure the information was useful, instructive, and easy to follow. Solar power is slowly growing around the world, and I hope to do my part in speeding up the process of making it a primary energy source. Please enjoy the content of this book and use it for reference as much as you want.

PERCENTAGE

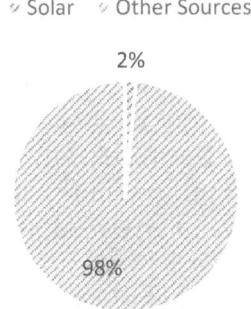

According to Solstice, about 1.6% of U.S. electricity generation comes from solar. It is growing yearly, but we can do much better than this. I hope this book entices many people to consider a switch over to solar power as an energy source. After reading about all of the advantages solar has to offer, more people will be inclined to utilize the power of the sun.

PERCENTAGE

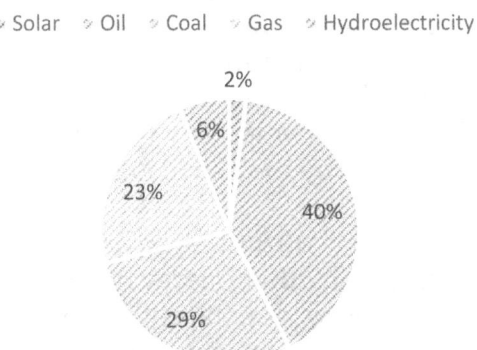

Compared to the rest of the world, about 2 percent of worldwide energy consumption comes from solar. The above graph compares it to the rest of the sources I will discuss below.

Chapter 1: Learning About Solar Power

Most of you have probably opened up your electric bill and nearly passed out after seeing the amount you owed. All of the appliances, the furnace, heater, and various products you have, suck up a lot of energy when combined, and your bank account can hurt a little because of it. Furthermore, there is countless research being done on the impact the other energy sources have on the environment.

Many experts have been pushing solar power as an alternative energy source, and many who have tried it are excited about the results—especially having lower energy bills. The target focus of this chapter will be to thoroughly detail what solar power is and how it is superior to other types of energy. Imagine saving money and saving the environment at the same time.

Common Sources of Energy

While solar power is slowly becoming more popular and attractive for people around the world, it is still nowhere near the most common source of energy. I will go over some of these in this section before discussing our main topic. Before that, though, let's define the terminology. Energy is the capability of doing work. Any type of movement, change, or shift, etc., is done by the use of energy.

The reason modern civilizations have been able to evolve is that they've learned to alter energy from one form to another. We use energy constantly during a 24 hours period, whether we are walking, driving our cars, cooking food on a stove, heating something in a microwave, and lighting up our homes. Energy has allowed us to sail ships across vast oceans, fly planes to any area of the world, and even shoot people into Outerspace.

I am willing to bet that most of you in the modern world are not fueling your stoves or furnaces by shoveling coal into them. Also, I have never seen anyone driving a car with their feet by pedaling their feet, like in the Flinstones. I know it's a cartoon, but still, there is some basis in reality here. There have been many advancements in automobiles over the decades, and a large part of that is the type of energy we use to move them. Just in the last couple of decades, everyday items have become much more convenient because they are easier to powerup and handle. The common forms of energy are:

- Heat
- Light
- Motion
- Electrical
- Chemical
- Gravitational

Energy can also be divided into two general forms.
- Kinetic: Working energy, or energy in motion
- Potential: Stored energy

Imagine a slingshot for a moment. When you pull the band back with a stone and hold it in place, that is potential energy. Once you let the band go and the stone shoots out, that is kinetic energy.

Fossil fuels can be dangerous

Energy has been around since the beginning of time in some form or another. The advancements have become exponential and will continue to grow. Perhaps unknown energy sources will poke their heads around the corner. Everything is evolving, and we are constantly finding new and safer ways to use energy to propel our lives forward.

Oil

Oil has historically been used as the world's most utilized energy source and is estimated to be about 39% of the total energy consumption. In the past, the percentage was higher but has declined over the past two decades. Nonetheless, it still has a stronghold worldwide. The main reason for the continued high demand is the emerging economies that rely on it. Over the year 2013, oil had the highest growth among fossil fuel types, with a gain of 0.8%.

China and the U.S. are the two largest consumers of oil in the world. Fossil fuels, like crude oil, have been a successful source of cheap, instant energy. These fuels are literally made from fossils, mainly in the form of dead plants that did not decay. As fossil fuels are burned, stored energy that was in the original fossilized organism is released. This energy was captured through photosynthesis over millions of years. The plants were captured from deep in the sea and converted to oil or gas products.

There is a huge chemical process that occurs, which we do not need to get into. Just know that fossil fuels store their energy until they interact with heat and oxygen. This is why a forest fire will go out of control on a hot day and in an open-air environment.

In addition to being cheap and reliable, oil is very well developed. This is because it has been used for so long, and many advancements have made it extremely efficient. It is also easy to use work that requires extreme energy, like driving a car or powering everything in a home.

The advantages make oil a very appealing energy source and the reason it is still the most popular in the world. However, the extreme disadvantages make finding an alternative source imperative. The following are a few of the downsides of using oil as an energy source.

- It contributes to climate change. Fossil fuels are not a green energy source and contain high amounts of carbon. Many environmentalists and experts on the subject have targeted oil and other fuels in this category as the main contributor to global warming. Much controversy still exists over all of this.
- It is a non-renewable energy source. The reserves are finite in amount, and they are not replenished naturally. Unlike other renewable sources of energy, oil formation takes millions of years to build to any noteworthy amount. If the resources for these fuel types erode, then it may not be the most practical and efficient source much longer.

- The oil reserves are being used up in an unsustainable manner. The silver lining here is that it forces us to look elsewhere for better energy utilization.
- Oil spills from accidents and natural disasters can wreak havoc on our oceans and sealife. These incidents are becoming much rarer, but even one spill can create turmoil for a long period of time. The oil and gasoline that comes from ships are not helpful either, as they are also being dumped into the ocean.

Will we be able to get rid of oil in our lifetime? Probably not; however, the quicker we can make the transition to other sources and learn how to use them efficiently, the more likely we can shift away from oil usage.

Gasoline is a major fossil fuel

Coal

Coal is another fast-growing fossil fuel around the world. It accounts for roughly 28% of the world's energy consumption. Coal consists of black or dark brownish sedimentary rock with heavy amounts of hydrocarbons and carbon. The substance is very combustible. Millions of years are needed to form and contain the energy stored by plants that lived in swampy forests hundreds of millions of years ago. Throughout millions of years, layers of dirt and rock covered these plants, creating an extreme amount of heat and pressure. Creating energy for any usable amount takes a long time, just like with oil.

There are many reasons why coal is an appealing energy source:

- It is the cheapest of all of the fossil fuels.
- It is a stable energy source, and most countries have their own plentiful supply of it. This means they don't have to rely on other nations for their resources.
- New strides and techniques have made coal mining and burning more environmentally friendly.
- There is a lot of employment available in the coal industry. This includes jobs in mining, transporting, and burning.

In addition, coal has been known to remove arsenic and other harmful substances from water. This means the water from the seas, oceans, lakes, or rivers that come into contact with it ends up being cleaner and more usable. Of course, there are disadvantages to coal as an energy source as well. It has not been put under the best light, and there are many reasons why.

- It is not a renewable energy source. Coal is another fossil fuel with finite resources. If it continues to be used as a common energy source, it will eventually

become depleted. We currently have centuries worth of stockpiles available, but they will not last forever.
- High levels of carbon dioxide are contained in coal, which is one of the largest sources of air pollution.
- Coal can produce very high levels of radiation from the coal ash, which is a byproduct of burning the substance. According to Scientific American, a coal powerplant can actually produce 100 times more radiation than a nuclear power plant. This produces both environmental dangers and health dangers for people.
- Coal emissions have resulted in more cases of asthma and other respiratory issues. Breathing in coal dust has been a link to lung cancer over time. Many coal miners who have been in the industry for decades are suffering from an issue known as black lung, and they can literally die of suffocation.
- Coal mines can impact green spaces, waterways, and other vital areas for habitat and cause dangerous amounts of pollution.

A growing percentage of the population wants coal to be done away with as an energy source because of the negative environmental and health effects it has.

Gas

Gas is the least utilized fossil fuel but still consumed by about 22% of the world's population as an energy source. Just like the previous fuel sources, it was formed well below the Earth's surface millions of years ago from the broken-down remains of ancient plants and animals. The decaying matter resulted in gas that was trapped inside porous solids that later became covered by harder rocks.

Methane, the lightest of the hydrocarbons, is what natural gas is primarily made up of. So, it emits lower levels of carbons monoxide, carbon dioxide, and nitrous oxides. It also does not produce harmful particles like ash, which can be detrimental to one's health. Hence, gas is heavily touted as an alternative, clean energy fuel source to oil and coal. Natural gas has no color or odor, which could lead to dangerous situations if there is a leak. The following are some of the common disadvantages of gas as an energy source.

- It is not a renewable energy source, so the supplies are limited. Once we use up the stockpile that we have from millions of years, we will be depleted of this resource.
- Even though the emissions are lower, natural gas does still produce carbon dioxide when burned, which can lead to environmental consequences, like climate change.
- Because of its highly flammable nature, any type of leak can lead to a huge explosion. The fact that it's colorless and odorless makes this danger even worse.
- While it can be transported easily through pipelines, the infrastructure is quite expensive.
- Fracking, which is a controversial method to extricate gas, has been linked to earthquakes.

What Toxins Can Fossil Fuels Send into the Air?

Hydroelectricity

Hydroelectricity is a fast-growing energy source, as it's usage makes up about 6% of the energy usage worldwide. Many Asian countries have steadily been moving towards this energy source. This energy source is created by generators that are pushed by the movement of water. It is generally created by dams that block water from the flow of a river. The water is collected or pumped into a reservoir. The high pressure behind the dam forces water through pipes that lead to turbines. From here, the water forces the turbines to turn, which leads to the generation of electricity.

Hydroelectricity is currently the most used renewable energy source in the world. It is environmentally friendly and does not pollute the water. Also, it is a highly reliable source of energy as there is very little deviation in the electrical power being created. The only time a fluctuation really occurs is when there are manmade changes. As long as there is water, energy will be created. In fact, adjusting water flow and output can alter the amount of electricity that is created if this is ever desired.

Compared to fossil fuels, this is a safe energy source that does not result in health consequences. Of course, there are some disadvantages that we must mention.

- Because of the shifting and damming of water, hydroelectric plants may impact many fish species due to the changes in water levels and velocity. Many of these species may lose their habitats and shelters.
- Building a hydroelectric plant is not cheap. On the other hand, once the plant is up and running, maintenance costs are very low.
- If there is a drought, then electricity generation can be affected, and energy prices can go up. Still, they should not be as high as with fossil fuel usage.
- Currently, hydroelectric reservoirs are in very limited supply.

Beyond some of the negative effects on fish, the environmental issues are practically nonexistent with this renewable energy source.

Nuclear Energy

Nuclear energy is about 4.4% of the world's energy source. Nuclear energy originated from the splitting of a uranium atom. This process, known as fission, generates heat to produce steam, which is then used by a turbine generator to produce electricity. Fuel-burning is not part of this process; therefore, no greenhouse gases are emitted.

Nuclear energy can provide power 24/7 with no carbon emissions. This makes it a clean energy source compared to fossil fuels. In addition, the energy density is higher when you split a uranium atom than when you burn fossil fuels. This makes the energy output much greater too. It is estimated that nuclear energy is about 8,000 times more efficient than fossil fuels. Of course, several dangers of this energy source do exist, including the following:

- Just like fossil fuels, uranium supply exists in limited quantities. Therefore, the reserves will not last forever, but they are expected to be around longer than all fossil fuels.
- Nuclear plants are expensive to build, but once they are operational, the costs to run them are pretty low.
- Nuclear power plant accidents are absolutely devastating. Two notable incidents are the disaster in Chernobyl, which set off huge amounts of radiation, and the crisis in Fukushima in 2011.
- On that same note, nuclear energy creates a lot of radioactivity, which can lead to many health consequences and fatalities. Radiation in small amounts is not that dangerous as we are exposed to them all the time. The radiation from nuclear power plants is anything but small.
- There are some environmental consequences of nuclear energy. This is mainly during the mining process of uranium, which leaves behind radioactive particles. These cause corrosion and can pollute nearby waterways. Underground mining can also be dangerous for exposed miners.
- This can create security threats. If uranium falls into the wrong hands, it can be used to produce nuclear weapons.

Geothermal Energy

Have you ever relaxed in a hot spring? If so, you know exactly what geothermal energy is. The core of the Earth is about as hot as the surface of the sun. This is because of the slow decay of radioactive material in rocks at the center of the planet. When deep wells are drilled, underground water that is very hot comes to the surface as a hydrothermal resource. This water is then pumped through a turbine to create electricity.

This is how geothermal plants work. These plants have low emissions if they pump the water and steam back into the reservoir. Geothermal energy is a reliable and renewable energy source that is available all year long. Once a geothermal system is set up, there is very little maintenance involved as the lifespan of the heat pumps is very long.

While geothermal energy is an environmentally friendly source, the risk of earthquakes can increase in geological hotspots. There is also a chance that specific locations will cool down over time, which means harvesting geothermal energy in the future will become very difficult.

Biomass Energy

This type of energy is produced from organic materials and used commonly throughout the world. The substance chlorophyll, which is present in plants, captures the energy of the sun by photosynthesis, which creates carbohydrates through the conversion of CO_2 from the air and water. When these same plants are burned, the water and CO_2 are released back into the atmosphere.

Unfortunately, this type of energy produces large amounts of CO_2 and not very friendly to the environment. In the absence of proper ventilation, air pollution can lead to major health concerns. Furthermore, inefficient use of biomass leads to the destruction of vegetation and degrades the environment heavily.

Wind Energy

You have probably seen those wind turbines all over the place, especially while driving through vastly open areas. Wind contributes to about 3% of the world's energy generation and has witnessed double-digit growth in consumption. As wind causes the turbines to move, they capture kinetic energy and use it to generate electricity. The electricity then travels to a power grid where utilities and power operators send it to where people need it.

Wind power is an effective energy source that has very little environmental impact, as it is also renewable. The wind will always exist. It has also been great for the economy in rural areas by creating over 100,000 new jobs over the years. Wind also does not come without its own disadvantages.

- The wind fluctuates; therefore, so does the energy. While wind will always be present, the speed changes,

and so does the spinning of the turbine. The counter this and make it as insignificant as possible, utility companies try to research the best locations to put up wind turbines for maximum efficiency.
- Wind turbines are expensive, but once they are up, operating costs are low.
- They pose a threat to wildlife, especially birds and bats. However, new research shows that the damage may not be as serious as with other structures, like cell phone and radio towers.
- Wind turbines are noisy and can often be heard from hundreds of meters away, depending on the direction of the wind.
- Many people don't find wind turbines to be visually appealing. They feel they block too much of the landscape. Of course, other people love seeing them.

Public acceptance of wind turbines is growing slowly as more wind farms are sprouting up

Since we have been moving towards the direction of renewable energy anyway, why not transition to the most underused source, which is solar power.

Describing Solar Power

We have officially arrived at the focal point of our discussion, which is solar power. This energy source is generated from the sun in the form of electrical and thermal energy. The sun is in the sky anyway, providing warmth anyway, so why not harness its potential in a much more efficient manner? Through a complex process called the photovoltaic effect, the energy from the rays of the sun is harnessed and converted into electricity to help power refrigerators, lights, the television, and other appliances in our homes. Solar energy can even power much larger buildings, cars, and numerous other items we rely on every day. In addition, small tools like calculators and watches can also work off of solar power, eliminating the need for excessive batteries or charging equipment.

For billions of years now, the sun has been producing an immense amount of energy. Its radiant energy has been powering life on Earth for millions of years. In addition to humans and animals, plants also rely on the power of the sun to generate the resources to live.

Unfortunately, solar power is one of the most underutilized sources of energy for everyday living, despite the fact that popularity and education have improved. We still have a long way to go in order to make this method one of, if not the main energy source to power up the world. The more we can learn to take advantage of its effects, the more exponentially the benefits will increase. If solar power continues on its upward trajectory, then we have the potential to live off the sun almost exclusively, without the need for other energy sources, especially the dangerous ones like fossil fuels and nuclear power.

Imagine how many disasters can be averted by reducing the need for these archaic energy sources.

The Benefits of Solar Power

Solar power provides many advantages that make it an ideal energy source. Many households are slowly switching over to renewable energy sources, and using the rays from the sun can certainly lead to this cleaner energy surge, as well. All life forms on Earth rely on the power of the sun in some way, shape, or form. Much more can be done to harness its unlimited power. The following are some of the benefits of solar power

Clean and Renewable:

It is 100% clean and renewable. It reduces the reliance on dangerous fossil fuels that have unnecessary health and environmental consequences. If we are to continue using oil, coal, and gas at a higher rate, our air, water, and soil will continue to be damaged, resulting in the loss of many species over several decades. It is estimated that more species can become extinct from fossil fuels between the years 2000 and 2065 than in all of the previous years of existence combined.

In contrast, solar produces no pollution, and the sun's abundant power offers an unlimited source of energy. It does no harm to the ozone layer or the Earth's landscape. Until the sun actually goes away, solar power is here to stay.

Freedom to Control Electricity:

Many households experience power outages on a regular basis. This includes developed countries, especially during natural disasters. The electrical grids in the U.S. are over 100 years old and are not built to handle the increased population size and extreme weather events. To upgrade the electrical grids, it will cost each state billions of dollars, and the fees will be passed onto the customers.

With continued technological advancements, we need to advance our infrastructure. When you start using solar power, you gain energy independence. You don't have to worry about the grid going down and halting operations. This provides a lot of peace of mind in knowing you won't be left in the dark, literally, without any electricity. When that next storm or disaster strikes, you will be ready.

Save Money:

The cost of home energy expenses are constantly going up and won't be slowing down anytime soon. Solar offers year-round efficiency and savings. Even during cloudy days, the sun still emits rays. So, don't believe that you will go without power when you can't see the sun. This could actually reduce your monthly energy bill all the way down to zero. Eliminating those expensive electrical bills alone makes this investment worth it.

In addition to reduced bills, you can get many tex benefits too. There are significant tax credits being offered on a regular basis, so check out which ones you qualify for. The cost of the equipment is getting cheaper, as well. I will discuss this more in a later chapter.

Why is Solar the Superior Renewable Energy Source

With all of the renewable energy sources that exist today, what makes solar power superior? While I respect any option for renewable energy, solar definitely stands on its own. Because the sun is always there, it is the most accessible to the general public. Let's compare some of the ways solar power will be the best energy source for you, compared to other top renewable energy options.

Solar and Wind:

Solar power provides many advantages over wind energy. First, when implemented on a smaller scale, solar is much cheaper and also more practical. A large wind turbine will not fit well into a small area. Solar power can also act more efficiently, especially when the sun is shining brightly. For a wind turbine to have an effect, the wind needs to be going at least 10 to 20-miles-per-hour.

There is no noise issue with solar panels. Private turbines can actually create more noise than large public ones. Since they are smaller, they can rotate at a much faster rate. This creates a higher decibel count. Wind turbines also need to be built to stand tall, so they catch the wind. Basically, unless you have a lot of acreages, wind energy is not practical. You will have to rely on public wind turbines and will not be able to control your energy amount like you can with solar power.

Wind turbines are also subject to more regulations. It is much more difficult to get approval to erect a wind turbine than it would to install solar panels. Solar panels and equipment are also much easier to install than the material needed for wind energy. Overall, solar is a superior energy method for your private home.

Solar and Hydroelectric:

Both of these methods are two time-tested forms of renewable energy. Compared to fossil fuels, solar and hydroelectric technologies are much more environmentally friendly. However, solar power slightly beats out hydroelectric in this regard. Solar power productions pose very few risks to the environment. Whatever environmental issues there are, come from the manufacturer and the materials used to build solar panels. The actual process of converting sunlight into a usable form of energy is very safe. Hydroelectric, on the other hand, has major effects on fish populations from the damming of the rivers.

Solar power also has an advantage when it comes to availability and access. The sun shines pretty much anywhere, and most places in the world get enough sunlight to generate electricity for the day. On the other hand, hydroelectric energy is limited to areas where the water supply is adequate. A certain water supply amount is needed to power the turbines and other generating equipment, and many areas of the world are excluded in this regard.

Wind and solar working together

Solar Power and Solar Energy

Solar power and solar energy are terms that are often used interchangeably; however, they technically mean something different. Solar energy is a more broad term. It refers to any amount of energy the sun radiates through visible light and other unseen electromagnetic waves. Solar energy systems refer to any technology used to convert the sun's energy into a usable form of energy.

Solar power is a division, or subcategory, of solar energy and more narrowly refers to the conversion of the sun's rays to electricity. The term I will be referring to more in this book will be solar power. Once again, the specific technology used for generating solar power will be discussed in a later chapter.

History of Solar Power

Since the sun has been around for billions of years, solar energy has been in existence since long before life on Earth even existed. As far as active use by humans, we were using solar energy since the seventh century B.C. when history details people using magnifying glass material to start fires. A few years later, the Greeks and Romans were known to harness solar power with mirrors to light torches that were used for various ceremonies. These mirrors became known as "burning mirrors." These mirrors are documented to be used by Chinese civilization for the same purpose during the year 20 A.D.

Many of the iconic Roman bathhouses typically had sunrooms that were facing the south side of a building. These sunrooms used massive windows to direct sunlight into concentrated areas. Many centuries later, during the 1200s, Pueblo Native Americans situated their abodes on cliffs facing the south to capture the sun's warmth during the cold winter months. People were understanding the benefits of the sun and taking full advantage.

During the 1700s and 1800s, researchers and scientists had more good luck with solar power by harnessing sunlight to power up ovens for long journeys. They were also able to develop solar-powered steamboats. It is safe to say that the concept of manipulating the power of the sun has been around for centuries.

With the advancements in technology came solar panels. There is much debate as to who deserves credit for these panels. Many people credit French scientist Edmond Becquerel, who determined that light could increase electricity generation when two metal electrodes were set into a conducting solution. This famously became known as the "photovoltaic effect." The photovoltaic effect was instrumental in the development of the element selenium, which has photoconductive potential. In 1876, scientists

Richard Evans and William Grylls discovered that selenium creates electricity when exposed to sunlight.

American inventor, Charles Fritts, produced the first solar cells made from selenium in 1883. This is why Fritts is considered the true inventor of os solar cells. However, modern-day solar cells are made of silicon, not selenium. The first silicon solar panel was created in Bell Labs in 1954. Bell Labs was originally founded by Alexander Graham Bell. Daryl Chapin, Calvin Fuller, and Gerald Pearson of Bell Labs are credited with the silicon solar panel.

A timeline of some other significant events are as follows:

- In the year 1958, the Vanguard I Satellite used a one-watt solar panel to power its radios while in outer space. In the same year, Vanguard II, Explorer III, and Sputnik-3 were all launched with some solar technology on board.
- In 1964, NASA was able to launch the first Nimbus Spacecraft, which was a satellite that could run entirely on a 470-watt solar array.
- In 1966, NASA made more progress by launching the first Orbiting Astronomical Observatory, which was powered by a one-kilo-watt array.
- Next came the year 1973, where the University of Delaware was responsible for building the first solar building. It ran on a hybrid model of solar thermal and polar photovoltaic power.
- In 1981, the first aircraft to run on solar power was built. It flew across the English Channel, starting in France and ending in the U.K.
- In 2016, the first solar-powered airplane, Solar Impulse 2, made its voyage around the world.

These are some of the major advancements in solar energy and power throughout the centuries, and we are continuing to grow at exponential rates. With the continued advancements and understanding of solar power, the sky is the limit, literally, with how much we can harness the sun's energy.

If we continue on the trend to utilize solar power as an energy source, it can soon replace fossil fuels completely and even provide energy more quickly and efficiently. People are woefully unaware of the amount of toxic waste we are injecting into the environment, whether it is the air, water, or soil. While I am not implying we should get rid of technology, if we can use it efficiently without harming the Earth, then why wouldn't we embrace solar power? The transition does not have to be overnight, but it should be consistent. People have understood the potential that can come from the sun for thousands of years, so not capitalizing on the advancements already made would be a mistake.

Innovative Solar Products

To give you perspective on how far solar power has come, I will go over some unique and innovative products that could change the future. The innovations are just getting started too. As a fun fact, NASA estimates that you would have to blow up 100 billion tons of dynamite every second of every day to mimic the same energy being produced by the sun. It is no mystery why scientists and researchers all over the world are trying to harness the amazing power of the sun. There is definitely more than enough to go around.

Solar Paint

Solar paint is applied to any surface that will capture the energy from the sun and convert it into electricity. It looks like typical paint, but it has billions of pieces of light-sensitive material that turns it into an energy-capturing paint. A technician can be installed by a technician who comes out after you have painted, which reduces the high cost of solar panel installation.

Solar pain is only about three to eight percent efficient, so it is not currently being utilized. To compare, a traditional solar panel functions at 18% efficiency. So, there are still some advances to be made.

Solar Windows

Scientists and researchers around the world are trying to advance solar windows. Many different engineers are developing various types of solar windows. For example, at the National Renewable Energy Laboratory, they are working on windows that transform from transparent glass to a tinted state. When in a tinted state, the windows convert sunlight into electricity.

Solar windows will never replace conventional solar panels because the windows must remain at least partially transparent. However, they can still produce a fairly large network of small photovoltaic sources.

Solar Cars

A lot of fossil fuels are currently being burned because people drive around so much. Our limited gas and oil resources are being used up quickly with the advent of more vehicles that are hitting the roads. Even less developed countries are expanding their use of automobiles, which further depletes our finite resources.

As a result, many engineers are trying to design solar-powered cars for consumer use. Fully-powered solar cars would give off zero emissions, but there are debates on whether they are realistic for public use. However, engineers are actively studying ways to make it work. Solar cars can already reach speeds of up to 55-62 mph in controlled environments.

Solar Roads

The future of solar transportation may be more than just vehicles, but the pathways they drive on, as well. A solar road initiative has been launched by many countries around the world. The first road of this kind in America debuted in 2016 in Sandpoint, Idaho. Engineers working on this project believe that if the U.S. used solar panels to cover the 48,000 square miles of paved surfaces in the contiguous 48 states, we would have three times more energy than the nation's needs.

Advocates of solar roads argue that they would drastically cut down on greenhouse gases. Solar panels used on solar roads can also be used to filter stormwater, melt snow, and light up the roads to give warnings to motorists.

With the potential that solar energy has, there is no doubt it can become prominent and even the main energy source for the entire world. We have a long way to go since it is only about two percent of the total energy source, but slowly, we can get where we need to be.

Chapter 2: Setbacks of Solar Power

Even the best things in life have their setbacks. The things we hold dear to us will let us down on occasion. No matter how much we love something in life, it will not be perfect. At least, not in all circumstances. The same holds true of solar power. Despite all of the positive qualities this energy source has, there are certain complications that we all must be aware of. At this time, it is not a perfect plan to rely completely on solar, even though I believe this will change in the future with further advancements. While I don't think the setbacks will warrant not going the solar route, I want all of you to be informed to make the best decision possible.

The Cost of Solar Panels

Solar panels have a significant upfront cost, and this can really turn people off from getting them. While the cost differs state-to-state and with different manufacturers, the average can run between $15,000 and $25,000. This is according to the Center for Sustainable Energy. This is definitely a major investment to start off with, and many individuals are not sure if it's worth it. Many people do not have this amount of money just lying around, and if they do, they certainly don't want to spend it all at once.

All I can say is that the initial investment will be high; however, the saving you get down the line will make up for it in no time. Saving a couple of hundred dollars on your electric bill every month can make the investment worth it within a couple of years. Let's say, for a moment, that your monthly electric bill is about $200, on average. In one year alone, you are spending $2,400. In 10 years, that is $24,000 already. If you take proper care of your solar panels, they will last much longer than ten years. I've got news for you. Your bill will only become higher from this point on if you choose to stay on the traditional route.

In addition to saving financially, you will be helping the environment in a major way. To make your decision simpler, here are a few tips to see if you will actually save money in the long run with solar panels.

Review Your Electric Bill

If you have a high monthly electric bill, then solar power can offset this expense or completely get rid of it. Even if your bill is not super high, you are probably still paying more than you should because of air conditioning on hot days and the heater on a cold day. However, the more your energy bill is, the better a solar panel investment will be. Multiple factors will affect your monthly dues, so assess what your average is and then determine how much you would save, even if you cut the bill down by half.

Many solar manufacturers even offer payment plans, so you don't have to pay for everything upfront. Pay a little bit each month, depending on the terms of the payment plan. However, even if you can separate it out into a few separate payments, it will still be a big help. Many companies will work with you if you are serious about switching over to solar.

Going Solar Can Save Money in the Long Run

Evaluate Your Sunlight Exposure

While solar panels can work anywhere, the sun shines, the more sun exposure you have, means more energy is being produced. Specific places in the world do better with solar power because of this reason. The orientation of your home, the type of roof, and the amount of shade produced also affects the solar system's output: the better your sun exposure, the more potential to save on energy bills. Some of this can be managed, like reducing the amount of shade from trees or placement of the panels. Other things, you can't control because, well, the sun loves to do its own things.

Estimate Residential Solar Panel Cost

The brunt of solar power expense comes with the purchase and installation of the actual panels. Most systems don't require much maintenance after the upfront costs and are designed to last for many years or decades without problems.

Their efficiency remains the same over this time. Talk to your neighbors and get estimates from at least three to five different contractors. Do your best to get the more beneficial deal possible. The only way for that to happen is by doing the proper research.

Look for Incentives

Homeowners can get several different incentives from the government for installing solar panels. For example, there is a federal tax credit that allows taxpayers to claim a certain percentage of the installation costs. This can significantly lower the tax amount you have to pay at the end of the year. Any additional credits vary by location. Certain states are more generous than others, so you have to see what the specifics are in your area. You can reference the database of state incentives for renewables and efficiency.

As solar products become cheaper, these incentives will go away. However, you are still saving money in one way or another.

The Option to Lease

You don't actually have to buy solar panels. There is also an option to lease, which will lower your upfront costs. You simply use the solar equipment for a certain period of time and make monthly payments while you have it. After you are done, you just return the equipment. Of course, you will not be eligible for incentives, and it will not raise the value of your home since you have to return all the equipment.

Other Drawback of Solar Power

If you decide that the solar route is right for you, then look at it as a lifelong investment. Before making this investment, consider the following setbacks.

Location Means A Lot

Your location on the globe is a major factor in determining the efficacy of solar power. Not all locations are created equal in this regard. If you do not have an adequate amount of direct sunlight in your area, the efficacy of solar will drop dramatically. This means that the more distant you are from the equator, the more challenging this process will become. Residents in places like Canada, Russia, Alaska, or Northern Europe are at a major disadvantage. On the other hand, places like Hawaii, Ecuador, and Tahiti are at a significant solar advantage.

The various seasons also determine the efficacy of solar power. Much more energy will be generated during the summer months than the winter months. Even though direct sunlight is needed for maximum efficiency, solar panels can still work if it's cloudy outside by generating power through direct sunlight. So, even if sunlight is partially or completely blocked by clouds, solar power can still be utilized.

Installation Areas

This is usually not a big ordeal for homeowners because the panels will be installed on rooftops. However, large companies that want to produce a lot of power will need a large installation area to provide needed energy for everyone on a consistent basis. The area has to be empty, with nothing else going on.

In Spain, there is a field that is about 173 acres filled with solar panels. They provide energy for 12,000 households. The field cannot be used for any other purpose.

Reliability

A common concern people have is how solar power works when the sun is not out. While it's true that electricity cannot be generated at night, this does not mean that all of your lights and appliances will suddenly shut off as soon as the sun goes down. At night, the energy that has been generated during the day and stored will get used up. It is to your advantage not to use more power than you need during the day, so more of it gets stored up. For example, you can rely more on natural light when the sun is out and never leave more lights on than you need. Many people are gone from their homes during the day, so they use minimal energy at this time.

Yes, reliability can be a big issue with solar power, but there are ways around it. If you absolutely need it, you can still connect to the main power grid at night, but then you will have another energy bill to take care of.

Associated Pollution

This may sound strange after touting the environmental benefits of solar up to this point, but there is associated pollution involved with solar panels. Mainly, there are some toxic materials and hazardous products used during the manufacturing process of photovoltaic systems. These can indirectly harm the environment. That being said, the pollution issue is minor at most and does not come close to other energy sources. I just want you to have all of the information you need.

Common Complaints About Solar Panels

While solar panels are relatively low-maintenance, they are not foolproof. There are still multiple problems that can develop. Therefore, if you notice your system acting below standards, act on it immediately before the problem becomes worse. The following are some issues to be aware of after you have installed these structures on your roof.

Snail Trails

Snail trail contamination occurs when brown lines show up on your panels, giving the appearance that snails moved along the surface. These spots may look innocent enough, but they might be indicative of underlying manufacturing issues and could create severe problems down the line.

Snail trails usually do not manifest for several years after purchasing panels and can be caused by several factors, including a defective silver paste. The silver paste is used during the manufacturing of the panels. This can eventually lead to oxidation caused by moisture. In addition, snail trails can lead to micro-cracks in the photovoltaic system, all of which can significantly reduce its functionality. The system can even fail prematurely. This is why it's important to assess your system on a regular basis. If you made the investment, you want it to last as long as possible. With proper care, the system can run for 20 or more years easily.

Birds

These fun creatures are great for flying around in the sky and singing to you while the sun is coming up. However, when they start making a mess on your car and other personal belongings, they are not so appealing anymore. This includes making a mess on your solar panels. Unfortunately, different birds can end up nesting underneath your solar panels and prevent them from doing their job properly. If you notice a lot of birds in your area, then consider installing some mesh wiring or other protective mechanism around your solar panels. These can protect your panels against other creatures that can climb on your roof, as well.

Inverter Issues

Luckily, the most expensive part of a solar power system is the panels, and they usually last a long time. The inverters, which are used to convert the energy, do not have the same lifespan. Expect the panels to last at least 20 years with proper care, but the inverters to start having issues about five to ten years earlier. The average solar user reports having to change their inverters every ten to 15 years.

Faulty Wiring

Faulty wiring can produce many electrical issues that will not allow your panels to perform as they should. Loose connections, corrosion, and oxidation are among the issues that interfere with electrical production. The wiring should not be mishandled, so be completely comfortable with it before installing your panels to avoid major issues in this area. Faulty wiring can be dangerous and create expensive problems down the line.

Roof Issues

Solar panels will cover one whole section of your roof and can protect it from the elements. While the solar power system will not hurt the integrity of your roof in most cases, the installation can certainly hurt affect it. If issues with your roof exist after the installation, make sure to get it taken care of. If you feel it is related to the installation process and someone else installed them for you, then let them know as soon as possible.

Since the roof is generally where homeowners install their panels, the next section will go over how to make sure your roof is up to par. This is imperative to assess prior to installing your system.

Make Sure Your Roof is Ready

Before buying and installing your solar power system, make sure your roof is ready and deal with any potential issues beforehand. Many considerations need to be made before placing your rooftop panels.

Does Your Roof Need Repairs?

Has it been some time since you climbed on top of your roof? In fact, have you ever done it? Many homeowners have never seen their roofs up close by climbing up there and taking a look for themselves. This means much degradation could have occurred without anyone realizing it. Before installing your solar panels, make sure that your roof does not need any repairs, especially in the area where you expect installation to occur. It is best to take care of any repairs, no matter how minor they may seem, before you actually go ahead with solar system installation.

You will have to deal with many more issues if you don't take care of the problem beforehand. For example, to get to your roof, you will have to dismantle all of your panels and put them up again after making the essential repairs. Why put yourself through this trouble if you don't have to?

What is the Shape of Your Roof?

Roofs are not all one size and come in many different shapes. Make sure that your roof will have enough space to place the number of solar panels needed to power your house. For example, if you need ten solar panels to adequately provide enough energy for everything in your home, but your roof can only handle eight, then it is not very practical to try to place these panels. Also, consider which direction the slope of your roof runs. Direction plays a huge role in this regard.

How Much Weight Can Your Roof Handle?

The last thing you want is for your rooftop to start caving in. While this likely won't happen, it is still important to know how much weight your roof can handle. Having your roof collapse on you is not only dangerous but very costly. Have your roof professionally evaluated to determine the weight capacity it has. Extra support might be required before installation. When you install the solar panels, make sure you are way below the weight capacity.

Where Will the Water Go?

Of course, your roof will collect a lot of water from the element. On a regular slanted roof, the water will fun down and collect in your gutters. It will not stay and collect on your roof. Once you install solar panels, the ancillary equipment like rackings and wire harnesses can prevent water from sliding off the roof as it should. Sometimes, solar equipment can even divert water in a different direction, causing it to pool on top of your roof. This can lead to damage and leaks.

Once the damage becomes severe, it will have to be repaired. The solar panels will have to be removed and reinstalled once again. When placing the panels on your roof, ensure they will not affect the runoff of the water.

While there are numerous setbacks to installing a solar power system, many of them can be worked around to prevent issues from occurring. With the right installation and maintenance, your solar power system will actually give you very little trouble in the long run. The benefits will usually outweigh the problems. If you have read about the potential problems and are still not deterred from going solar, then keep reading.

Chapter 3: The Initial Process

Now that we have gone over what solar power is and how it compares to other energy sources, I hope you are ready to make the investment to help lower your bills and make the world a more environmentally friendly place. Imagine having to worry less about gas explosions, radiation poisoning, or nuclear disasters. The sun is way up high, staring down on us every day. It continues to emit an infinite amount of energy that powers life on Earth. Let's start concentrating that energy the sun gives off and make it as useful for us as possible. I can't think of more natural ways to create sustainable and usable energy for the world.

Solar Power Equipment

Setting up your home to be solar can be an arduous task at first because of all the equipment involved. I will dedicate this section to discussing the different materials needed for this process. The main goal of going this energy route is to reduce costs in the long run and minimize our carbon footprint. Use this as motivation to finally transition into powering your home with the energy from the sun.

No matter what design and products you ultimately get, the functions of this system will generally be the same. The solar power system you install will work by absorbing natural sunlight and converting the photons into a usable energy form. These types of systems are often referred to as photovoltaic or P.V., solar power systems because of the scientific process that occurs.

Installing these types of systems can reduce the need for a community power grid that supplies electricity to cool, heat, light, and operate your home. As a result, you will have a clean, renewable energy source with little maintenance after the initial installation. Once your solar system is paid off, you will have a free energy source for years and maybe even decades.

In more urban areas, like residential neighborhoods, solar power systems use a grid-tied method. This means they are still connected to the main power grid that can be initiated for back up energy. This way, once your solar reserves run out at night, you will still have energy from another source. Yes, you will receive an energy bill, but it will still be significantly lower than using the power grid 24/7. So, you will not have to worry about losing power in the middle of the night while watching a long movie or doing some extra work in your living room.

On the other hand, in more rural areas with underdeveloped land, the homes will have to rely on an off-grid system because there will be no main power grid. As a result, they will have to rely on solar alone once they choose to go this route. I will not get into the main components of a solar power system.

Home Off-Grid

Solar Panels

The panels are the main component of any solar system. They are what absorb the energy given off from the sun. They will then convert the particles of energy in this light, known as photons, into usable electricity to power electrical loads. These panels can be used in multiple places, including remote areas like cabins, as well as residential and commercial buildings.

Solar panels are comprised of individuals solar cells, which are composed of many layers of silicon, phosphorus, and boron. The phosphorus provides a slightly negative charge, while the boron provides a slightly positive charge needed to make the reactions for change to occur. These reactions, which are quite lengthy and complicated, are known as the Photovoltaic Effect, which was mentioned earlier in this book.

The average home has more than enough area on its roof to fit the necessary number of panels needed. The panels will generate more than enough energy to power up the hoe while the sun is out. The excess energy produced will go into the main power grid of the home and then be used for electricity at night.

Solar panels are a very efficient way to produce electricity for various applications. Many people who want to live off the grid love the idea of producing solar power through this method because they don't have to rely on anyone else but the good old sun to provide them energy.

Solar panels in a field. Often times done for shared or community solar energy resources.

The Solar Inverter

This is the hardest working piece of equipment in the solar power system. Its main function is the convert the direct current flowing from the solar panels into an alternating current to use for your home. The other functions of the inverter include:

- Voltage tracking
- Grid communication
- Emergency shutoff

For the voltage tracking function, the inverters will continually track the solar array's voltage to determine the maximum power at which the solar panels will work. This will ensure the system creates the most power at all times. You will need two different inverters for grid-tied and off-grid systems.

The grid communication function makes sure that no power from the solar panels makes it outside your home in the event of a temporary power outage. This prevents line workers who are troubleshooting outside from getting zapped. The inverter will also feed power loads into the grid when power is not needed by your home and the battery is full. This is if the grid is connected to your solar power system.

When a hazardous electrical arc is detected, which is caused by system aging and degradation of the material, inverters will shut down for safety reasons.

There are four different types of inverters that function in their own unique manner.

String inverters:

These are used for wiring panels together. There are many stringing configurations that will impact how each system will perform. These types of inverters are meant to pair with sets of panels that are simply strung together. When multiple panels are connected like this, the production output is reliant on the performance of the worst producing solar panel. So, if one of them is shaded, the whole set will be affected. If your roof has good southern exposure, then this won't be much of an issue. The great thing about string inverters is that they are the most affordable.

Microinverters:

These types of inverters perform the job of converting direct current to alternating current at the back of every panel in the system. This solves the problem of having a few panels that are compromised by shade. Even in constantly shady conditions, maximum levels of the alternating current shift from the solar panel into your home and the main grid. Research done at the University of Virginia reported 27% more efficiency in partially shaded installations with the use of microinverters. Having an inverter on individual panels can allow for monitoring of unexpected performance problems.

Solar microinverters are more costly than string inverters, but they allow for easier system expansion.

Power Optimizers:

These are similar to microinverters in how they are affixed to the back of each solar panel. Once again, this allows for individual panel monitoring. The difference is, they do not convert electricity from direct current to alternating current. They actually track the voltage of the direct current flowing through the solar array's strings to ensure optimal production gets sent to the inverter. The optimized direct current then gets sent to a smaller inverter to convert to alternating current. Power optimizers are great for battery backup systems. These setups are highly efficient and more affordable than microinverters. The power from your solar panels can charge the battery directly, which avoids system losses due to conversion from direct to alternating current and back to direct again.

Hybrid Inverters:

These inverters pair well with home battery backup systems too. They have the ability to convert direct current from the solar panels to alternating current for your home and also the capability to convert alternating current from the grid to direct current used to charge your battery bank.

Hybrid inverters also come with charge controllers that can detect when to direct solar energy to your home, the grid, or the battery. Also, it can detect when to pull electricity from the main grid to the battery. Many hybrid inverters are more affordable upfront.

Solar inverters are the reason the electricity from your solar panels becomes usable in your home. Get multiple estimates to determine which kind of inverter is best for you.

Solar Racking

Solar panels are not directly attached to your roof. They are mounted on a proper racking system, which also allows for creating a proper angle to get optimal sun exposure. Good solar racks are essential to make sure your expensive panels are installed properly. Not having proper racks is like having an expensive car with really bad wheels. These rackings are usually made from aluminum, which works well for rooftop installations. Aluminum is strong, durable, and does not weigh much.

Solar Performance Monitoring

The solar performance monitor will show how much electricity is being generated per hour, per day, and per year. This is to verify the performance of the photovoltaic system. This monitoring system can also detect potential performance changes. This tool is needed to verify if your particular solar power system is operating at its very best. You can even detect problems before they become a major issue.

The solar performance monitor operates through the inverters. As your inverter converts direct current to alternating current, information about the power levels and production is sent to various cloud-based monitoring systems. Homeowners can then access this information through a smart device or through some type of mobile app. It's amazing how we have been able to control the sun to a degree through the internet.

If a system has a power optimizer, then it won't rely on a wireless connection to transmit data. Therefore, you can continue to monitor your system if there is an internet outage.

Solar Storage

Also known as the solar batteries for energy storage to use later or overnight. The battery basically works as they should and provides energy for a solar power system to operate when the electrical grid is not available. So, while the system is collecting energy from the sun during the day, any excess amounts are transferred to the battery. At night, when sunlight is not available, you can still power up your home and appliances by using the energy left in the battery. Be mindful that this will be in limited amounts until the sun rises again.

The following are some of the different types of solar batteries to consider:

Lead Acid:

These are some of the oldest types of batteries and are tried and tested. These batteries can last anywhere between two to eight years, depending on how often you discharge them. Their cycle life is between 1000-3000. If you go for these types of batteries, it is best to store them in a shed away from the heat. Heat can significantly reduce these battery's lifespan. These batteries only have a discharge capacity of 60%, which means you can only use 60% of their potential. Any more, and it degrades the battery much more quickly.

Lithium-ion:

These are the type of batteries you have in a laptop or iPhone. These have a longer life than lead acid batteries, with 4,000-6,000 cycles. Plus, they have a discharge capacity of 80%. As a result, they last about 13-18 years on average. The main drawback is that they are about 50% more expensive than lead batteries.

Flow:

A flow battery is made up of a water-based solution called zinc bromide that flows between two tanks. During charging, zinc is extracted from the liquid and stored separately. During discharge, the zinc is put back into the liquid. The zinc flows from the big plastic at the bottom to the electrodes on the top. These batteries have 100% discharge capacity, and this does not affect the lifespan of the battery. The following are a few more advantages:

- They can withstand a high level of heat.
- The zinc-bromide solution is a natural fire retardant.
- Because of the physical separation of the battery's components, there is no chance of an explosion.
- They can sit on a shelf without being charged and not degrade.

The main disadvantage here is that they have a lifespan of about 4,000 cycles, maximum, which is slightly less than lithium-ion batteries, on average.

Sodium Nickel Chloride:

These batteries have a broad operating temperature range between negative 4-Degrees Farhenheit and 140-Degrees Fahrenheit. They are also fully rechargeable with no toxic or dangerous chemicals. Finally, there is no fire risk due to the chemistry of the battery.

The main disadvantage is that the expected lifespan is about 3,500 cycles at 80% discharge capacity. These batteries are also much more expensive than lithium-ion batteries, running about $20,000 installed.

Determining the Number of Solar Panels Needed

To determine how many solar panels you will need for your home, you must first figure out some information beforehand. Mainly, you will need to understand your goals as far as investment, lowering your carbon footprint, and saving money in the long run. Remember that the more you rely on solar, the larger your initial investment will be, but the more savings you will have in the long run. To calculate how many solar panels your home will need, you will have to consider the following criteria.

How Much Solar Power Will You Use?

You have to figure out your home's average energy requirements to determine your solar power needs. Look at your past utility bills to find this information. From here, multiply your hourly energy requirement by the peak sunlight hours in your area, then divide this number by the specific panel's wattage. Roof size and battery storage also play a role here.

Solar Panels fitting on a rooftop with plenty of surface area

How Many Watts of Energy Do You Currently Use?

On your energy bill, you will see a section that says, "Kilowatt-hours used," or some similar phrase. Also, not the time period, which is generally 30 days, especially for a monthly bill. Finally, look at your daily average. If your bill does not have a daily average, then divide your monthly average by 30 or yearly average by 365. To determine the hourly average, if not listed, divide the daily average by 24 to get your final answer. In turn, if you have an hourly average but no daily average, multiply the hourly by 24.

Whatever your daily energy usage is in kilowatts is what your target daily target for solar energy should be. This will cover most of your energy needs. Experts recommend adding an extra 25% cushion to your target daily average since solar panels do not work at maximum capacity at all times due to weather issues and other factors. So, if your daily average is 30 kilowatts, 25% of this is 7.5, so account for 37.5 kilowatts daily needs.

How Much Sunlight Can You Expect?

The Peak sunlight hours for your specific location will have an impact on the energy you can expect your solar power system to create. An area with fewer peak sunlight hours would require more panels to obtain the same results. For example, a home in Seattle would require more panels than a home in Phoenix or Maui.

Going back to question one, multiply your hourly usage by 1,000 to calculate watts. Divide the hourly average of wattage by the number of peak sunlight hours in your area. This final number will give you the amount of energy your panels, when combined, need to produce every hour, on average.

What Affects Solar Panel Efficiency?

The efficiency of a solar panel refers to how well it can convert sunlight into usable energy. Not all solar panels are alike, and photovoltaic panels, which are the most commonly used types for residential areas, can range from 150 to 370 watts per panel. Cell technology also plays a role here, as solar cells with no grid lines on the front absorb more sunlight than conventional cells. In addition, a microinverter on each panel can optimize power conversion over a single mounted inverter on the side of the house.

The key takeaway here is that the more efficient the panels, the more wattage they can produce. In turn, the fewer panels you will need. This is where quality over quantity plays a role. If you are trying to get cheaper panels to save money, realize that you might need to buy more of them to gain the same energy outcomes.

To determine the number of solar panels you need, divide your home's hourly wattage that you calculated earlier by the wattage of the solar panels.

Does Solar Panel Size Play a Role?

The size of a solar panel can certainly make a difference. However, just because a panel is bigger, it does not mean it is better. The more important number to look at is the power rating. A comparable example would be batteries. A 9-volt battery is smaller than a D battery, but it still produces more power. The same can be true for solar panels. Also, as technology has advanced, solar panels have become more efficient while also becoming smaller in physical size.

If you have a larger usable roof area, buying more larger panels at a lower cost and reduced efficiency might be the way to go. If you have a limited roof area, then smaller, more expensive, and higher efficiency panels might be the way to go. In short, size does make a difference, but not in the way you might think.

After you answer the above questions, you will have a better idea of how many solar panels are ideal for your situation. There is no one-size-fits-all answer here.

What's With the South?

I have mentioned facing the south a few times in this book. This is the optimal direction your solar panels should be facing for maximum efficacy of the system. We have been told countless times that the sun rises in the east and sets in the west. This fact has become common knowledge; however, it is not completely true, which I will explain in my first point. The bottom line is, you want the most out of your solar power system, so you cannot just buy any old panel and plop it wherever you choose. You will be wasting both money and the efficiency of the product. I will go over some major reasons why the panels you install should be facing towards the south. There is a lot more to it than just leaving the panels in the sun.

Why is Direction Important?

The idea of the sun rising in the direction of the east and setting in the west is only 100% if you live directly on the equator. Otherwise, it actually has a Southern offset, especially in the winter. The sun also does not travel in a straight line as the day progresses. It arches a little bit as it moves. To get a good visual of this, watch some time-lapse videos on the internet.

This is why solar panels need to face southwards.

What About the Angles?

On top of the direction, the angle of the solar panels matters too. There is some math involved when installing the panels. The proper angle is determined by the geographical latitude of the given location. This will produce the most amount of energy in a given year.

Direction is Important

Protecting Your Equipment

If you bought a car, a piece of jewelry, or a new T.V., I imagine you would protect these items at all costs. The same should hold true for your solar equipment. You will make a significant investment in these materials, and while they are built to last for a while, you need to do your best to protect them and increase their longevity.

Remember that your solar panels will be out in the elements, and mother nature can be unforgiving at times. While you cannot predict or control what will happen in your environment, there are certain steps you can take to keep your valuable equipment safe and secure. I will go over some of those in this section.

Protecting Solar Panels from Hail

Hail can do an immense amount of damage to cars, homes, and other items caught in their path. They can range in size and are sometimes get bigger than golf balls. Unfortunately, solar panels are not immune to the barrage of a hailstorm. The good news is most manufacturers understand the issues that hail creates and specifically design their panels to eliminate the risk of damage as much as possible. Any reputable brand has tested their products multiple times. Of course, what happens in a controlled environment, versus out in the elements can be two totally different things.

While modern-day solar panels are nearly hail-proof, there are some extra precautions you can take to minimize damages even further. If you are investing thousands of dollars into a new system, you might as well take as many precautions as possible, just as you would with any other expensive item.

Add a Layer of Methylcrylate:

This is a type of additional type of armor or plate made of plastic to help shield the panels further and make them more durable. A thin layer should be applied to the most fragile areas of a panel and be done as carefully as possible. Make sure not to block the access line that leads to the power supply.

The higher the quality of your plates, the more impact they will be able to handle. You may have to spend more money but will provide extra protection for your solar panels.

In lieu of methylcrylate, you can assemble a piece of plexiglass that is one inch larger than the dimensions of the panel. Assemble the plexiglass on top of each panel with a small space between each of the layers. This will provide an ample amount of armor.

Once Again, Angles Matter:

Another viable option to protect against hail is the mount your solar panels on a pole and then tilt them into a vertical position when a hail forecast is predicted. This way, most of the hailstones will slide down the side of the panels rather than hit them directly.

Track Weather Reports:

Keeping up to date with the weather can keep you informed about any impending hail storms that are on the horizon. This goes beyond just watching the weather channel. You can also get weather apps on your phone or go down to your local weather station to become fully informed. From the weather station, you can get a detailed report that notifies you about any serious weather attacks that are in the forecast.

Assessing the weather report to be prepared for any potential hail storms or conditions

Add Solar Panels to Your Insurance:

As soon as you get solar panels and equipment, make sure they are covered by your homeowners' insurance as soon as possible. If they are not, get them added. This type of insurance can cover much of the damage that occurs to your solar power system. Also, be aware of your warranty information. Find out how long it lasts and what exactly is covered.

Most solar panels nowadays can withstand hailstones up to one inch in size and at speeds of up to 50 miles per hour. Therefore, they can definitely withstand a lot of impact. Most hail storms will not have much of an effect, but it's still good to take as many precautions as possible.

Damage from Falling Debris

There is a lot of debris that falls from the sky, and much of it gets on our roofs. If you have ever cleaned out a gutter, then you know precisely what I mean. Debris can range from small items, like leaves and small twigs, or larger and more damaging materials like rocks and tree branches. While solar panels are durable, they can still be damaged by the garbage flying through the air. This is especially true over time when we go long periods without checking on our system.

Mostly, the smaller debris causes micro-scratches on the solar panels. These scratches may not seem like much, but they can significantly reduce the absorption of solar energy because the sunlight will not shine directly on the solar cells. In turn, this will decrease the energy output to your home. Larger debris can actually break panels completely.

While there is no way to eliminate all of the debris that exists, a few extra precautions can be taken to reduce its effects. The most obvious solution is to make sure there are no trees or other structures directly over the solar panels from where objects can fall. If this option is not possible, then at least maintain the upkeep on your trees by keeping them pruned. This way, you won't have to choose between your trees and solar panels, and trees provide many benefits.

Furthermore, do not allow the debris that does get on your roof to build up. This will significantly reduce the efficiency of your panels and lower the energy output to your home. If you clean up the smaller debris using a microfiber cloth and garden hose, then it will keep your solar power system clean. Doing this once a season is good enough, but you can choose to do it more often, as well. If you live in an area where debris flies through the area constantly, then monthly cleanings on a minimum might be ideal. Basically, if your solar panels look dirty, then clean them.

Water Damage to the Panels

Water damage to solar panels generally occurs because the seals have become old and deteriorated. This is similar to the insulation or sealing of window panes. As this sealant around the solar panels gets older, it becomes less effective and allows water to leak through. This can eventually lead to short-circuiting of the components in the solar panels. In order to avoid seal damage, take the time to reseal your panels on a regular basis.

Protection From Thieves

Since solar panels are expensive, they are an invite for robbers and thieves to come to take them while they are unguarded. When installing your solar power system, it is important to take precautions for preventing theft. The following are some ways to protect your solar panels.

- Use an alarm system. Thieves can easily dismantle your solar panels within a few minutes. So you can go to bed one night, or leave for work in the morning, and then find some of your panels missing the next time you see them. You can attach an alarm system to your panels that will go off with any movement. You will receive a notification on your smartphone or to your alarm provider.
- Have motion detector lights on your roof near your panels that can alert you to any suspicious activity.
- Take note of the serial number on each panel. In the case of theft, authorities can use these numbers to help them track down your panels.
- Link the solar panels together. Items are much stronger in a group, so if you link panels together, it

will be difficult to remove one easily and hard to take a whole bunch without getting caught. In most cases, a thief will give up and run away quickly.

Once you start buying solar equipment, you will understand the value and will want to protect it as much as possible. If you are willing to make the investment, take whatever precautions you can to keep it working for years to come.

Buying Used Solar Panels?

In many cases, if people want to pay less for a car, they will get a slightly used one and hopefully have it checked out to learn the history. The same holds true for houses and other items. When buying something that is used, all I can say is, "buyer beware!"

Solar panels are best when they are bought new. They will be at their most efficient stage and will usually have some type of warranty. Of course, many people cannot get over the initial investment, no matter how much money they can potentially save down the line. In this case, they may opt for used solar panels to reduce upfront costs. However, just like anything else used, there are potential pitfalls to be aware of. If you decide to go this route and shop for used solar panels, then the following are a few things to watch out for.

Lower Energy Output

The older solar panels get, the more they lose their ability to output power. This is especially true if the panels use amorphous silicon. While this fact alone should not prevent you from buying used solar panels, just be aware that you might need more of them to match the efficacy of new ones. For example, you may need ten used panels to equal the benefits of eight new ones. Again, this depends on the exact age of the solar panels. If you are planning on buying extra equipment that is old to match the value of new equipment, then it might behoove you to just buy new. Crunch the numbers to see where the real savings are.

Loose Connections

If you don't look closely enough, you could miss loose connections between the solar cells when buying used panels. These loose connections can significantly reduce the absorption of sunlight. You can fix this issue by tightening the connections with the proper tools, but make sure the bargain you are getting is worth the extra trouble.

Damaged Panels

Solar panels can take a lot of damage over time. If you notice cracks in the glass, moisture underneath the cover, or completely broken connections, then determine if you are willing to fix the issues. If so, the bargain might still be worth it.

Unsteady Current

Always test your panels for their efficacy. Used solar panels can often be damaged under the surface, where it's difficult to detect. Often time, moisture can get into the internal circuitry, which can cause fluctuations in the voltage output. If you notice damaged seals, assume there is some internal damage, as well. Bad circuitry implies that the solar panels will not work.

Many people sell their solar panels because they want to upgrade to new ones. If you are ready to do the proper assessments, then the best places to buys used solar panels are sites like Ebay. Always inspect the panels thoroughly and never let anyone rush you into doing so. If they do, this should be a huge red flag.

The focus of this chapter was the equipment necessary for a functioning solar power system. In the next chapter, I will cover step-by-step instructions on how to install your panels and ancillary equipment.

Chapter 4: Installation

So far, I have discussed the benefits of solar power and the advantage it has over other energy sources, especially fossil fuels. I have also provided a detailed list of equipment needed and what their general functions are. The focus of this chapter will be to provide in-depth instructions on how to properly install your solar panels for safety and efficacy. Your initial cost will be large, so pay close attention to the installation process so you can maximize your investment. How you install the equipment is just as important as the materials that you buy.

Should You Hire a Pro

After paying a hefty sum for all of your solar equipment, the last thing you probably want to do is hire a worker to install it for you. This will set you back even further but may actually be worth it. There are many technicians out there who are specially trained to install solar power systems. They will pay attention to the little details and make sure to position everything properly for optimal use. You are dealing with sophisticated materials in relation to solar power. Therefore, hiring a pro could be to your benefit.

Even if you are good at construction, solar power systems are a different animal with different strategies. Consider some of the following reasons for hiring a licensed solar technician and then determine if you want to go at it alone or not.

Right Method of Installation

I mentioned earlier how the proper positioning and angles are needed for the optimal function of your solar panels. This practice will also help keep your expensive panels safe. Furthermore, your roof is a delicate structure that can be damaged quickly if not handled properly. Since most solar panels will be installed on the roof, this can become a major concern. Repairing a roof can become quite expensive, and the last thing you want is to add this bill after you just bought your solar equipment.

A solar installation contractor will analyze the best method to mount the panels on your roof and the best rackings to use.

License and Permits

The main advantage of hiring a solar installation contractor is the fact they know all the proper procedures to install them. They understand the specific requirements of the state. Furthermore, getting permits for the installation requires a lot of paperwork, which can be very time-consuming and give you headaches. It can be even worse if you are part of a home-owners association. A solar installer can do much of the legwork for you to make your life easier. Usually, you will just have to verify things.

Manuals and Precautionary Measures

The solar kits will come with many different instructions and manuals. If you have had difficulty installing simple appliances, then this task might feel completely daunting for you. If you make a mistake or become confused during the installation process, you could risk messing up the whole system.

A solar installation contractor has a wealth of training and experience in setting up these systems. They will also have all of the right tools and accessories. If any issues do come up during the installation process, they will be able to troubleshoot them easily.

Electrical Hazards

Electricity can be a dangerous element if not handled properly, and you will be dealing with plenty of it with a solar power system. If you are not familiar with the wiring and electrical structure of your home, then it can be dangerous to install one of these systems on your own. Before you even attempt this, become familiar with your electrical structure. If you don't want to take this risk, then definitely hire a professional to do it for you. There is no sense in putting yourself and your home in danger to save a few bucks.

Maintenance and Warranty

If you install the system yourself or perform follow-up maintenance work, then you may lose out on your warranty. With a contractor, you will have peace of mind in knowing that the work is guaranteed. If there is a screwup, the contractor of their company will take care of it.

The bottom line is, you must decide how much time and effort you want to put into installing a solar power system. If you are busy with work and family life, then this might be the last thing you want to do. Unless you really enjoy it. If you do have the time and want to install the system yourself, then keep reading for step-by-step instructions or proper connection.

Professional installation of solar panels is recommended

Getting the Proper Permits

Before you can begin installing solar panels, you will likely need permission to do so. There are various governing bodies out there that set the rules for different states, cities, and even neighborhoods. If you are unaware of the rules in your particular area, then do your research to determine what you are and are not allowed to do. It will become a complete headache to buy and install your equipment, only to be told later that you are not allowed to. To prevent you from breaking any rules, I will go over some quick steps to get you permits done right.

Research Your Local Regulations

Most city, state, and federal governments need homeowners to obtain permits before going solar. Research your local regulations and determine whether you need a permit or not. This can vary, but in the general sense, more rural areas will not require as much permitting as more residential and metropolitan areas. Crowded urban and suburban locations usually have more legal obstacles to contend with.

Determine What Documents You Need

Once you have informed your local government or regulation board of your plan, you need to start gathering all of the documents you need for permits. Here is a list of what is generally required:

- Permit application
- Owner-Builder Verification form
- Site Plan
- Roof Plan
- Construction Plan
- Elevation Plan
- Equipment Plan
- Electric Plan
- Location Plan

In addition to these forms, you will need:

- A Board of Zoning Appeal
- Shade tree commission approval
- Land development approvals (Land, sewer, water, fire, and flood)

Even if you have an installer take care of these for you, it is still a good idea to be informed.

Keep the Government Involved

While the government and the agencies involved are not always pleasurable to deal with, keeping them involved during the solar process is essential. The paperwork should include all the details about the inspection. Always keep these documents with you in case of an inspection.

You can save a lot of time and energy if you do the whole permitting process online. You can send emails instead of talking on the phone or going into the office. Hopefully, your local regulatory agency or government allows this.

Installing the System

Are you enjoying this book? I really hope to know your thoughts, I would be grateful thank you!
Now if you are ready to take the dive and have all of the proper permits, then it's time to start installing your system. Hopefully, you have bought all of your necessary equipment. While all of this material is quite durable, make sure to take as many precautions as possible. Even if you do plan on hiring a professional installer, understanding these steps can benefit you in order to stay informed.

Step 1: Finding the Best Location

Location! Location! Location! Where you physically put your panels is imperative. Once you have your supplies, survey your property closely to identify the best area to install your panels. Remember, it is not just about what area of your roof gets hit the most by the sun. The direction and pitch of the roof play a role too. All of these factors will determine which location will provide your panels with maximum exposure during most of the day.

Step 2: Building the Platform

After determining the location of your panels, you need to pick a suitable platform. These platforms can be flat or slightly tilted towards the sun. If you have a good, solid roof, this should save you the trouble of building a custom platform. I mentioned this before, but make sure your roof receives the maintenance it needs before starting your installation process.

If your roof alone is not suitable for solar panels, at the very least, you will need four concrete pillars and several pieces of 4x4 wood. The amount of wood depends on how many panels you will have. The conduit is to protect your wiring and other fragile material that needs to have a secure connection from the platform to your house. Once again, if your roof is your platform, a shorter conduit is generally required.

Step 3: Mount the Panels

Many solar panels come with mounting brackets that you can easily attach to your platform, especially if it's made of wood. For steel platforms or roofs, you will need a good drill. If your panels did not come with brackets, inquire with the manufacturers if they can be provided or where the best place to get them is. Also, ask them how to install the brackets on your solar panels.

Once you are ready to mount, secure the front feet of the brackets first. Make sure they are completely squared and centered prior to securing them with screws or other mounting material. After obtaining meteorological data to determine the movements of the sun in your area, it is time to attach the rear legs. By the way, I am aware that the Earth moves and not the sun, but for the sake of these instructions, bear with me.

Once you have these measurements, use the rear legs to elevate and tilt the solar panels in the best direction and angle. This will ensure the panels catch the most amount of sunlight. Make adjustments to your rear legs based on your calculations. After this, fasten the legs to your panel. You can make adjustments later on, but it's better to get things right the first time.

Step 4: Proper Wiring

This may be the most important section of the installation process, and I cannot stress enough the importance of being familiar with how the wiring works. This is true, even if you don't do the wiring yourself, so you can troubleshoot any performance lapses or other issues.

You will need a junction connector or a fuse combiner box. You will need to wire the solar panels to the conduit. The wires of the panels should be stripped and wired in pairs prior to being connected to the fuse box mounted on the platform. The stripped wire must be connected to the terminal box. The red wires connect to the positive terminal, and the black wires connect to the negative terminal. After connection, the wires are fed through the bottom of the junction box and connected again to the corresponding positive and negative terminal blocks.

The solar breaker needs to be connected to a circuit breaker disconnect. The energy will flow from the circuit disconnect to a charge controller and then be stored in a battery bank.

When needed, energy flows from the batteries back to the disconnect. From here, it flows between the disconnect to the power inverter, which will change the power from direct current to alternating current. After this, the alternating current will flow into your home's electric panel.

After gaining an understanding of how the wiring works for solar panels, it is time to connect them to the house. First, run the electrical cable from the solar array into the house through the underground conduit. If you connect the cables to a nylon rope, you can easily thread them through the conduit to the inverter panels. Connect these cables at the fuse combiner box that is located at the base of the solar array. The green cables are connected first to the grounding strip, the red cables are connected to the positive terminal block, and the black cables are connected to the negative terminal block.

Step 5: Ground the Panels and the Mounting System

It is now time to set up the grounding to the solar array. This ensures the panels are safe for human touch. To proceed, bury a grounding rod with about six inches of its body sticking out of the ground. This can be done on a wood surface or directly on the ground. After this, run a copper wire from the rod to the fuse box, then from the solar panels to the fuse box, as well. This will ground the panels safely so you can continue with the installation process.

Step 6: Connect the Electrical Components

After the grounding process, it is time to set up the home connections. Go back inside the building to complete the wiring. Run the cables from the solar array to the inverter panel disconnect. Once again, red goes to the positive terminal, green goes to the grounding terminal, and black goes to the negative terminal.

You can decide where to install the battery bank. It is recommended, though, to put it near the control panels. From here, run the conduit through the ceiling into the inverter panel. Feed two heavy-duty battery cables into the panel disconnect. The opposite ends of the cable will be attached to the battery pack. The battery system must be attached, alternating between positive and negative, and secured properly in place. After completing this process, cover the panels to protect them. You will now have a functioning solar power system. From here, you just have to do some test runs on the system to make sure it is functioning properly. Switch on the power and see how the system is working.

If you are installing a grid-tied system, the utility company may require a dedicated meter to measure the energy production from the solar array. In most cases, you will just have to build the base, while the utility company will install the meter face.

I admit that the installation process is not easy. It can be physically demanding and intricate at times. Be very careful and take your time to make sure you do not mess anything up. The last thing you want is to make a huge investment in your panels only to mess up during installation. If you don't feel comfortable, then definitely don't hesitate to hire a professional. It will be worth the investment in the long run. A professional will also make sure all of the regulatory aspects of a solar power system are taken care of.

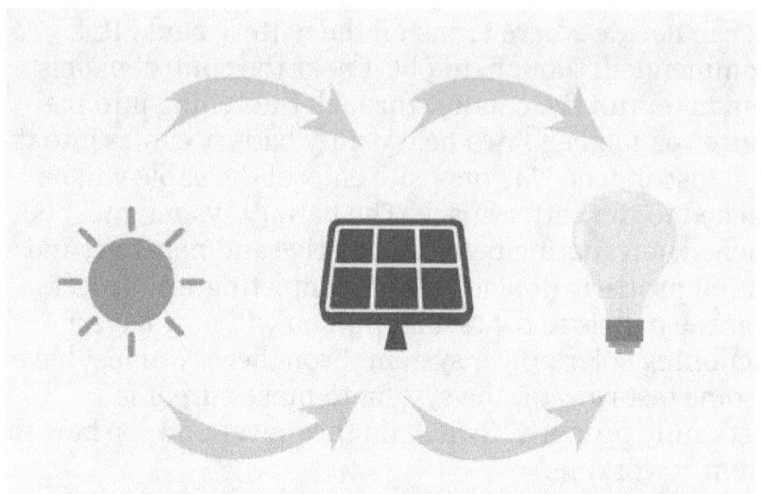

The basics of solar energy: The sun, the panel, the receiver

Troubleshooting and Continued Maintenance

Once your solar panels are installed, you want to make sure they are functioning at their maximum capacity. You will not wear out the sun, so take as much advantage of it as possible. If you are having issues, which is rare, if the installation was done well, then there are a few areas to look at.

The Wiring is Loose

Loose wiring can be the cause of many electrical issues. The wiring is what connects all of the components of a system. As a result, there are many areas where connections might be loose. You can use specific meters to help pinpoint the problem or retrace your wiring to find any faulty attachments.

The System is Overheating

If your panels are overheating because of excessive temperatures in the higher 90s and above, you might start to see some heat fade. Your panels can underperform because of this. If overheating is an underlying problem, you may notice less power during heat waves. This is a problem that should be addressed before installation. Make sure to purchase panels that are much more heat resistant. Many of these panels have a thin layer of material that sandwiches the microcrystalline layer, which makes them more efficient than other panels during high temperatures.

The System is Dirty or Damaged

The most common performance issue is the result of your system collecting dirt and other debris. Mother nature can be a panel's best friend, or worst enemy, depending on the way you look at it. It pays to hose off your panels at least two to three times per year. If you notice extra debris collecting, then you can increase the frequency as needed.

You can use a soft push broom in certain areas with hardened dirt. In addition, you can hire a cleaning crew. When cleaning a solar panel:

- Never use an abrasive sponge or soap as they may scratch the glass.
- Avoid harsh cleaning materials.
- Be very careful when getting on the roof as it can become slippery while washing your panels down with a hose.
- Consider buying a solar panel cleaning kit that comes with all of the supplies you need.

If you notice small cracks or damage to the panels, it should be addressed immediately. The functionality of the panel will be okay for a while, but the damage will eventually increase and make the system operate worse. A damaged panel, which is rare, needs to be changed out. If a particular section of a panel can be replaced, then that's also an option, depending on what the damage actually is.

The following are some more tips for proper maintenance.

- Keep panels out of the shade. Trim any nearby trees if you have to.
- Keep an eye on the inverters and make sure they are always flashing green. If they are not, then you are losing money by no longer substituting for your electricity use.
- Always Document and keep track of the system's day-to-day performance. Write down how much energy has been produced at a set time every day. Make special notes on days when there is inclement weather. The manufacturer will be able to provide you with the proper monitoring system for your panels.
- You can install automatic cleaners, kind of like a sprinkler system, if you don't have time to clean your panels by hand. These will go off based on a timer.

Luckily, solar panels do not have any moving parts that can be affected by rust. As a result, maintenance requirements are low.

What About in the Ground

Up to this point, I have been talking about installing solar panels on the roof, which is the ideal location for residential properties. However, solar panels are often mounted into the ground when it comes to commercial properties or solar farms. This option also exists for residential homes, as well. In fact, in some cases, it might be the ideal choice. For example, your roof might be completely covered by shade. Your roof also might not be made of material the will allow solar panels to be mounted properly. Finally, your roof may not have adequate space. There can be a wealth of reasons why solar panels will not fit on somebody's roof. The following are some of the pros of having solar panels mounted into the ground instead of on a rooftop.

- Perfect sunlight alignment. Since the ideal position for solar panels is facing south, your roof may not always provide this option for you. If your roof does

not have a south-facing pitch or gets a lot of shade during the day, then you might need more panels to make up for the lack of sun contact. Ground-mounted panels can eliminate the issue because they can be pointed in almost any direction for optimal sun exposure.
- If you own a lot of land area that is basking in the sunshine, then you can make great use of it by having your own little solar farm. Most people's lawns provide more space than their roofs. This means you can install a fairly large system to produce great amounts of energy. Just like before, you will need to get permits and check your local zoning regulations.

Now that I have gone over the pros, let's look at some of the cons:

- Ground-mounted panels are not anchored to an already secure structure, like your roof. Therefore, a secure structure needs to be built before you can install your solar panels. To do this, lay down a concrete foundation. This process can add extra time and money to your investment.
- Ground-mounted solar panels are not that appealing to the eyes when in a residential location. For a ground-mounted system to be efficient, the panels need to stay several feet above the ground and be set at an angle. There is nothing discrete and no way to cover them up. This can become an eye-sore for many people, and if you decide to sell your home later, the potential owners may not enjoy having a solar farm in their yard.
- The panels will take up most of your yard and block you from doing recreational activities.

In summary, the roof is still the best location for solar panels when it comes to residential property, but ground-mounted panels do not have to be out of the question

Ground-mounting as an alternative to roof-mounting

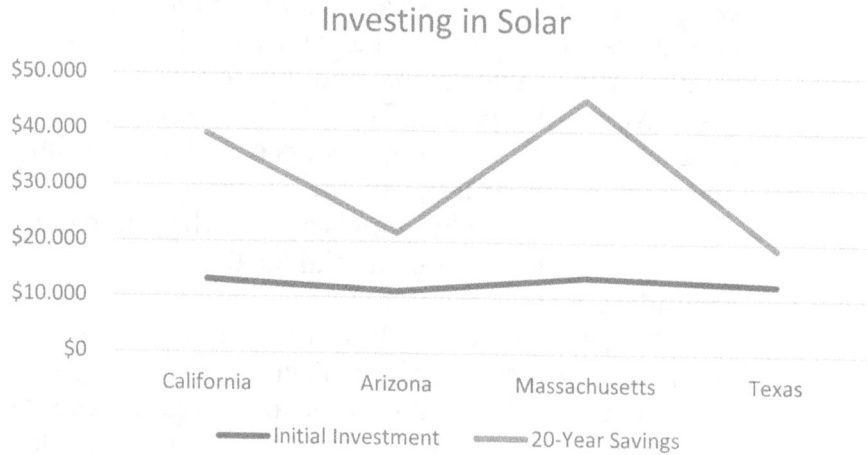

This line graph illustrates the average initial investment made and the 20-year savings across four different states. As you can see, the 20-year savings is much higher than the initial investment, showing that switching to solar power will save you in the long run. Properly maintained solar panels could last a good 25 years or more.

Chapter 5: The Best Products

This chapter might sound like an advertisement to a certain degree, but I will go over some of the best manufacturers and products in the solar world so you can buy the best that is out there. Much of the modern-day solar equipment is above standard and improving by the day. Solar power will become much easier to harness as time goes on.

Finding Who's the Best

It can be challenging to find a good solar company to trust. Solar power setups are not cheap, so you want to make sure you do things right the first time. I will go over some fundamental criteria when choosing your manufacturer, so you can make the best decision possible.

The first thing to consider is the history of the manufacturer. With the increasing popularity of solar, the number of new manufacturers is regularly increasing. These new companies are also offering cutting edge technology and promotional deals. However, all of this can just be a coverup. It is important to go with an established company with a good history. Also, make sure they are offering a minimum 25-year warranty on their products. If they are not willing to give this guarantee, they don't really stand behind their products. Choose a company with quality service that does not hurt your investment.

You must also look into the power and wattages, which are the basic features of a solar panel. Some manufacturers produce smaller panels with higher capacity. The larger panels will provide more surface area but less wattage. With the smaller panels, your installation costs will be lower too.

You must receive good customer service. This is not exclusive to the initial investment and installation, but during the follow-up maintenance, as well.

Finally, the price of the solar panels depends on the manufacturer and the quality of the products. Make sure to compare prices and what you are getting for those costs. This will ensure you get the best product for the best price.

Some Questions to Ask

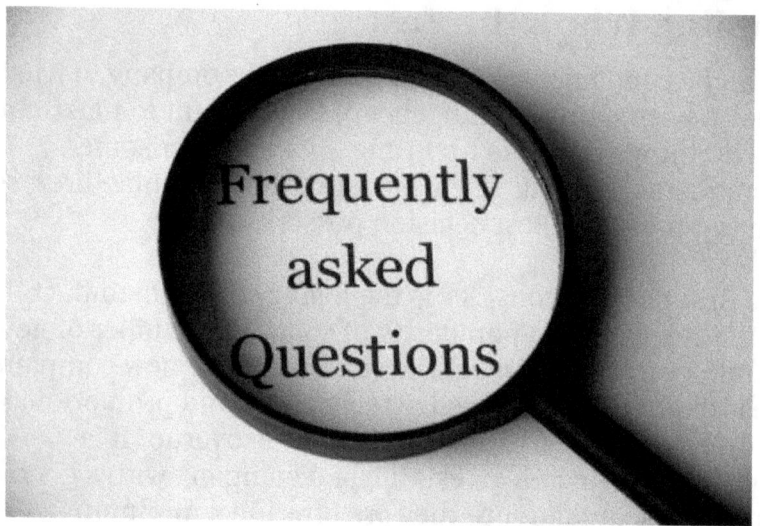

To understand your solar company and what you are getting yourself into, you need to know the right questions to ask. Never jump into a major investment like this until you are comfortable. Any reputable solar manufacturer will welcome your questions and do the best to educate you. The following are some important questions you need to ask, no matter what company you are dealing with.

How Much Money Will You Save Over 20 Years?

You should expect your solar panels to last a minimum of 20 years with proper installation and maintenance. Therefore, asking how much you will save during this time is a good question to start with. Of course, this will be an approximation.

How Much Will You Pay Upfront?

The payment structures will vary significantly with different solar offers. You might pay everything up front, set up a payment plan, or lease your panels. Whatever the case, determine what you will pay on the day of purchase.

Will the System Increase Property Value?

Numerous studies have shown that a solar power system will increase the value of most homes. Of course, there can be multiple variables at play, and your distributor certainly cannot give you real estate advice. If you lease your panels, then they will definitely not increase your property's value because you do not own them.

Questions for the Installer

The following questions are specifically for the person installing the solar panels for you.

- Do they have the necessary business licenses and insurance policies?
- What is their license number? (This is to verify their credentials).
- Do they install everything themselves, or do they have subcontractors? If they have subcontractors, are they properly licensed, as well?
- Will there be an expert electrician on-site when the system is installed?
- How long have they been in business?
- How many solar systems have they installed?
- How much experience do they have working with the utility company?

- Do they have customer reviews, testimonials, or references that can vouch for their work?
- What rating do they have with the Better Business Bureau?
- What type of guarantee do they provide for their work?

Questions to Ask About the Warranty and Replacement Procedures

- Who is the point of contact if there is a problem with the system? Where are they located?
- Who is responsible for making sure warranties are addressed in a timely manner?
- How many different warranties are there, and what exactly does each one cover?
- Does the warranty cover any damage done to the roof?
- What is the process of removing and reinstalling the solar panels if the roof needs to be attended to?

Questions Regarding Liability and Insurance

- Is there a possibility of a lien being taken out on your property? This can occur for the nonpayment of solar panels.
- Is there an additional insurance policy that comes with the system, or should you contact your homeowner's insurance agent?

As you begin shopping around for solar panels, these are some important questions to ask.

The Best Solar Manufacturers

The newer solar panels that are being created are more efficient and less hazardous to the environment. New metals and materials are always being incorporated into these panels, and most of these companies are competing for the best product. That being said, the following are some of the top players in the industry, and you can not really go wrong with any of them.

Auxin Solar

This is a solar manufacturing company that started up in 2008. The main headquarters are situated in San Jose, California. Their services are not exclusive to the United States as they serve international customers, as well. This US-based manufacturer produces solar panels, mounting racks, portable solar panels, tracker systems, and complete solar energy systems.

Auxin Solar is trying to develop commercial and residential products in Europe, Australia, and America. They have a wide range of services and have been providing quality products for 12 years.

Heliene

Heliene is a joint venture between North American and European investors. It is one of the most renowned solar companies based on the goals and services provided. Their objective is to establish a diverse solar energy production system around the globe. With two continents backing up the company, it definitely has a lot of leverage. This also means the products and services are top-notch.

This US-based company manufactures premium photovoltaic cells, durable solar panels, solar modules and delivers them all over the world. It is considered a Tier 1 company in the US for high-quality products. The price list might be large at the beginning, but you will save a lot of money down the line with all of their services.

Seraphim Solar USA

Seraphim is a highly ranked solar panel manufacturer in the US. It is an extension of the Seraphim group out of China. Their American headquarters are in Texas, while their manufacturing is done in Mississippi. The Company creates photovoltaic modules for residential and commercial purposes. In addition, they produce solar panels for government, industrial, military, and utility markets.

Seraphim is used by multiples sectors, which shows its value. They have a wealth of experience, too, as they've installed over six million solar modules in over 30 countries worldwide since its inception in 2011.

Solaria

This is considered a top solar panel's manufacturer in America based on its production capacity and service. This company has been around since 2000 and has over 65 patents that are registered for manufacturing, applications, products, and equipment. In addition to solar panels, the company provides automation, solar energy conversion, and estimation.

While the headquarters are in California, the company operates worldwide. Based on IP portfolios, this company is considered the most respected in the solar industry. Its customers receive outstanding service at an affordable price. This is why Solaria has a global reputation.

SunPark Technology

SunPark has over 20 years of experience in the solar industry and is a growing solar panel manufacturer in America. The parent company of SunPark is Yiheng Science and Technology Co. Ltd, which is based in China.

This company produces solar panels, PV cells, and other ancillary equipment for solar energy production. They offer great products for commercial and residential energy solutions. They also follow an eco-friendly production process while maintaining the quality of their products.

SunPark also stay current on the latest technologies, so they can provide them for their customers.

Solar World Americas

Solar World Americas has been in operation for 45 years, starting up in 1975. It is also the largest solar operating and manufacturing company in the entire Western hemisphere. They create grid-tied solar systems, residential solar systems, and commercial solar systems for small and large companies. In regard to service duration, it is the most experienced company in the world.

Their affordable pricing options also make them incomparable in the industry. They are one of the leading manufacturers of solar products because of their domestic and global customer satisfaction rates.

Panasonic

Two giants in the technology world, Panasonic and Tesla, started their partnership in 2018 to produce solar energy. The headquarters for this joint venture is located in New York, and the Panasonic solar panels are produced in a factory in Buffalo. Their high-efficient energy cells can produce more than 19.7% of the energy from the sun.

Panasonic has created the most reliable solar panels for residential and commercial usage, and they developed a complete system with Tesla Powerwall for off-grid solar installments.

The eco-friendly manufacturing and global operations make them a leader in solar energy. Their solar panels come with the latest technology and are always in high demand.

JinkoSolar

JinkoSolar is one of the top American solar equipment manufacturing businesses in the world. It produces and distributes several solar products in the country. It boasts a wide range of products, like solar panels, PV modules, batteries, off-grid solar power systems, and grid-tied solar power systems. They have operations in the UK, South Africa, India, United Arab Emirates, Spain, Italy, and Chile. The main advantage of JinkoSolar is that you will work with a worldwide service center and manufacturing units. You will receive efficient service wherever you are.

First Solar

First Solar is a leading solar panel and energy manufacturer in America. The company owns the world's largest grid-connected photovoltaic powerplants and invest a huge amount into research and development of advanced technologies.

The most significant achievement for the company is the stable grid integration system. They work to produce solar panels, equipment, complete solar solutions, and develop solar technology. Their solar technology is also cost-competitive. Their multiple operations enhance their service area and help include them at the top of the list of great solar companies.

LG Solar USA

LG Solar sells a wide range of electronic products all over the world. They are now producing solar panels and working to develop solar technologies. They have already developed all-weather solar panels that can work in any weather condition. These panels are also eco-friendly and come with advanced solar technology.

LG offers different types of solar products for industrial and residential purposes. They have a highly-rated solar-powered air conditioner, which is considered one of the best solar devices in the world. It offers high efficiency and low energy consumption.

JS Solar

JS solar started its operation in 2007. The company is located on the Southeast side of China, and they offer both engineering and manufacturing of solar products. JS produces solar panels for both industry and residences to provide efficient energy production services. They are also up-to-date on the latest technology the industry has to offer. The average solar panel provides 18% efficiency, but the JS models offer 22% efficiency. It is one of the highest energy generation rates for photovoltaic solar modules.

Canadian Solar

Canadian Solar Company was established in 2001 with a share of Chinese investments. The company is a manufacturer of solar products, plus engages in research for the industry. The products Canadian Solar provides are panels, modules, and mounting devices. Their modules come with high-performance design and are very efficient. For example, the Ku Modules have an 18.6% efficiency and max power of 370 wattages.

I know I provided a lot of random companies in this section, but these solar manufacturers are considered some of the best of the best in the industry. If you go with them, you will receive great service and state-of-the-art products.

The Best Solar Panels
The focus of this section will be the top solar panel brands based on a number of different factors. I will break down these factors into different sections. The information here is provided by EnergySage.

Highest Efficiency Solar Panels

A solar panel's efficiency depends on how well it can convert sunlight into usable energy. The following are the top manufacturers in this regard.

- SunPower: 22.8%
- LG: 22%
- JS solar: 22%
- REC Group: 21.7%
- CSUN: 21.2%
- Solaria: 20.5%

Solar Panels With the Best Temperature Coefficient

One of the unique qualities of solar panels is that they can handle extreme weather conditions and temperature. This is essential since they are out in the elements all the time. Based on the lowest temperature coefficient in a panel, the following are the top solar panel manufacturers:

- Panasonic
- REC Group
- Solartech Universal
- SunPower
- LG

Material's Warranty

Finally, we get to the material's warranty, which protects against equipment failure due to defects. The following companies all have warranties of this kind for 25 years.

- LG
- Panasonic
- Solaria
- SunPower
- Mission Solar

I hope this chapter gives you an idea of where you can turn to for your solar needs when you decide to go this route. There are many reputable companies and solar brands out there. The ones provided in this chapter are just a few examples to help get you started. Overall, you should understand what to look for and the right questions to ask.

Chapter 6: What if You Move?

Just like solar panels can be installed in one location, they can be removed and transferred to another area. Therefore, if you decide to move, you can take your solar panels with you. Of course, this can come with its own set of challenges. Moving solar panels and equipment is not an easy process, so make sure you do the proper research before you start dismantling.

Before Making the Move

Before you decide to move your solar panels, there are a few things you need to consider to make sure it will be worth it. This can be a grueling process, so it's important to make sure you are taking everything into account before you make a move. Make sure you are still taking advantage of all the benefits.

Are You Selling Your Property?

If you are selling your current home and moving to a new one, then it is natural to want to take your solar panels with you. You invested heavily in them, after all. However, understand that your home's property value can go up significantly, and selling for a higher price can offset the cost of your solar panels. Before putting your home on the market, calculate the potential return you can have and determine if it's even worth it to remove your panels. It may be easier to leave them there and install new ones when you get to your next house.

Once you sell the property with the solar panels still intact, then they are no longer yours to keep. Consider beforehand if you will be selling your house with the solar panels attached or not.

Location and Logistics

The new location you are moving to plays a major factor in whether you should move your solar panels. If you are moving somewhere locally, contact the company that delivered and/or installed your equipment and find out if they will move it for you. This is a better option than transporting all of it yourself or depending on a general moving company.

If having the original installers or manufacturers moving the equipment is not an option, then it might be to your benefit to leave it at your current home and just try to sell it for a higher price. Removing and reinstalling solar panels is an involved process. Many companies may not even remove them because of warranty concerns. A more practical solution can be to buy new panels for your new home.

Rules and Regulations of New Location

Contact the electric company of your new location to find out their rules and regulations for solar panels, especially in regard to reinstallments. This is where you will learn about zoning and permit requirements, as well. Basically, before moving your solar panels to your new home, make sure you are allowed to do so.

Sun Availability

Obviously, a major factor in deciding to move to solar was the availability of sun exposure in the area. Before moving to your new location, determine if the availability will be equal, or at least close. Always consider the different weather conditions and geographical locations before moving your solar panels.

Potential Damage

Even though solar equipment is quite durable, the potential for damage is always there. When moving your equipment, always take this into account. A lot of skill and experience is needed to move solar panels, so be as careful as possible. If the solar panels are over ten years old, then it probably is not worth it to move them.

Be prepared for damage to the roof, as well. You will notice that the area under the panels will retain their color, while the area outside of the edges of the panel will be discolored due to natural wear because of direct sunlight. In addition, as panels are removed, small holes will be left in the roofing material. If you are planning to sell your home and take your panels with you, these specific issues will need to be addressed with the buyer. They could expect you to repair the roof if they are to purchase the home after the removal of the panels. This will be another expense that might not be worth it.

Consider all of these dilemmas before deciding to move your solar equipment. Once again, selling it all with the house might be more practical, time-wise, and economically.

Safe Moving Practices

If you are set on moving your solar panels and do not want to leave them behind, then realize that the entire removal and reinstallation process can take about three to four months, counting all of the paperwork that is involved. The following are some basic steps for dismantling an existing solar power system. Before starting, get on those safety goggles and gloves.

- Take some large opaque cloths to completely cover all the solar panels. This will cut off the sunlight, and the panels will stop producing electricity.
- Using your voltage meter, set it to measure the direct current. Touch the metal part of the red positive probe of the meter to the positive terminal of the solar panel. Hold the metal portion of the black negative probe to the negative terminal of the solar panel.
- The meter should be reading zero volts. Do not proceed with dismantling the system until it does; otherwise, you risk electric shock.
- Disconnect the wires attaching the solar panels to the solar panel system. Loosen the bolts or screws while holding the wires in place.
- Some systems will have an extra solar power disconnect for added safety in case one of the panels becomes exposed to the sunlight.
- Cover the ends of each wire with electrical tape or a rubber terminal cover. This prevents physical contact in case the wires become live. Cover the solar panel wire connect terminals too.
- From here, you can carefully remove the equipment and detach the solar panels, including the platforms and conduit.

As you can see, it is not as easy as just taking everything apart quickly. The same level of care needs to go into dismantling a system as installing one. This is for the safety of the equipment and yourself.

Proper Transportation

Once you take down your solar equipment, then it's time to transport to your new location. Adequate care needs to be taken during the transportation process to prevent any damage to your solar materials, especially the panels. The following is a list of supplies you will need to package your solar panels safely.

- Palette: Use the original one you received when the solar panels were first delivered to you. This will offer sufficient stability and load capacity for the panels.
- Corrugated cardboard underlayment: To place on top of the palette.
- Stacking device: Four per panel.
- Foam pads: Six pieces per panel.
- Edge protector: Eight pieces per pack.
- Tension belts and straps: Four pieces of each.
- Cover foil: One piece with dimensions of about 2,000,1,300x0.04mm.
- Adhesive tape with pictograms for safety labeling.
- Cardboard protection pyramid to prevent stacking.

After gathering all of your equipment, follow these packaging steps:

- Check your palette for any obvious damage before using it.
- Place your palette on a flat and stable surface. Consider the necessary space needed for handling and final transport.
- Center the corrugated cardboard underlay on the palette. Make sure it is lying flush.
- Carefully place the solar panel with the front side, which is the side that faces the sun, towards the palette on the cardboard underlay. Make sure it is positioned centrally on the palette and not sticking out on the edges in any way.

- Take the six foam pads and place two on each long side and one on each short side.
- Position one stacking device on each corner edge of the panel.
- Continue stacking the remainder of the panels in the same manner. You can stack up to 30 panels on one palette.
- Position the top palette with the front side up.
- Place the edge protectors, two pieces per package corner.
- Place two straps at the long and short sides close to the palette outer blocks.
- Apply the straps as tightly as you can without putting too much stress on and damaging the panels.
- Cover and wrap the package using the material in the list provided above.
- Label the packaging with the following labels:
    - Fragile
    - Keep dry
    - Do not stack
    - Do not tilt
- You can also just use the tape with the pictograms on it.
- Wrap the entire panel packaging with the foil. It is recommended to wrap around 35 times on all sides for adequate thickness.
- Bond the cover foil to the packaging with the adhesive tape. Again, this is the tape with the pictograms of the transport instructions.
- Bond the cardboard protection pyramid to the top of the package.

From here, the package of solar panels is ready to be transported. Just like before, if you have reservations about doing this safely, then I advise hiring the proper movers who are experienced in transporting solar equipment. If you cannot find someone, then it may behoove you to leave the equipment behind. Once you are at your new location, it will be time to reinstall the panels again after getting all of the necessary permits.

Warranties

With the installation, movement, and maintenance issues that will arise, it is important to have a general idea of how the warranties on solar panels work. You want to avoid getting blindsided as much as possible. The following are some general guidelines for coverage that is provided on solar panels. Of course, you want to do your own research when you buy solar panels. Remember that if you buy used panels or move with them, the warranties could become void. Make sure you find out about all of these coverage details.

Manufacturer's Warranty

There are two specific manufacturer's warranties: A performance warranty and a product warranty. The performance warranty guarantees that the panels will not degrade below a specified level. This ensures that the equipment will continue to produce the needed power during the warranty period, which usually ranges between 25 to 30 years. The panels should still work at about 79-87% of the original performance.

The manufacturer does not guarantee a certain amount of solar production for your system. There are many variables at play here, so the performance warranty cannot guarantee this aspect. However, you will be assured that each panel can produce a specific amount of instantaneous power on its own.

For example, if your solar panel is in an area with adequate sunlight and was installed properly but still does not produce the desired output, then there might be something functionally wrong. The solar panels could be faulty, which would be covered under the performance warranty.

The product warranty is in place to provide protection in the case of defective materials or issues with workmanship during the manufacturing process. Per their discretion, the manufacturer can repair the damage or replace the whole product.

The product warranty does not cover the cost of labor to diagnose problems with a product. The cost of labor to replace the equipment or shipment of new products will also not be covered. Product warranties generally last between 15-25 years.

As far as the other equipment, the breakdown goes as follows:

- Inverters come with a ten to a 25-year product warranty.
- Batteries come with a warranty of about five to ten years.
- Racking comes with a ten to a 20-year warranty on workmanship and defects in the material.

Solar Installer Guarantee

Some solar installation companies will offer additional protection for the services they provide. Not only will these cover workmanship, but they can also guarantee total solar production of your system. This goes well beyond a performance warranty. A solar installer guarantee will make sure your product is actually performing and not just capable of performing.

A Solar Panel Warranty Becoming Voided

A solar power warranty can become void, so make sure you understand the rules to not allow this to happen. If you lose your warranty too early, then you are on the hook for a major expense without getting full use of your solar panels. The following are a few reasons why your warranties can become void:

- Not having them professionally installed. Having work done by a contractor who is not industry certified can nullify a performance warranty. The manufacturer may not be able to tell if it's faulty equipment or poor installation. If someone does install your panels for you, make sure to ask about a solar installation guarantee.
- Failing to maintain your solar panels adequately can lead to warranty problems, as well. For example, failing to keep branches trimmed or not getting rid of debris can affect your performance warranty.

Ask your manufacturer and any contractor you hire to give you a list of things that can void the warranty. Remember that your homeowner's insurance will also cover some repairs, so make sure to update your insurance agent/company as soon as you buy solar equipment.

One last recommendation I will make is to make sure you will be staying at your current location for a long time before installing solar panels. This way, you will get the full use of them. Of course, if you are a good investor, perhaps you can use them to increase the sale and rental values of your property.

Chapter 7: Who Does Solar the Best?

When it comes to solar power, all places are not created equal. Geographical benefits definitely exist when trying to use solar as an energy source. While you can still use solar energy and benefit from it if you live in places like Seattle, or other areas with less peak sun times, there are still specific locations around the world where solar power will be at its best. The focus of this final chapter will be to discuss the locations in the world that are currently benefiting the most from solar energy. Hopefully, other places will learn from them and get on board with this amazing renewable energy source.

Which Countries Do it the Best?

Several countries from around the world have been able to harness the power of that large fireball in the sky. Many environmentalists have made the claim that if most countries used the same amount of resources as the United States, then we would need two or three more planets to help keep up. While fossil fuels will run out at some point, the sun is not going anywhere for now. The entire world can benefit from sunlight, and it will not reduce the efficacy of this giant star. The following countries are proving that solar power is the alternative energy answer to fossil fuels.

Germany

Germany has been at the forefront for a long time when it comes to the production of solar power from solar energy. It produces roughly 38.2 gigawatts of the 177 gigawatts produced worldwide as of the year 2014. One gigawatt is roughly the output of a large natural gas or nuclear plant. In some instances, Germany has met close to 50% of the country's energy needs from solar power. It is the largest economy to rely on renewable energy.

Germany has a goal to rely 100% on renewable energy by the year 2050, completely eliminating the need for fossil fuels. The nation is definitely on its way by adding to its solar capacity every day.

China

China is the most populated country in the world and leaves the biggest carbon footprint. The fact that they are committed to renewable energy I definitely encouraging. The more they can utilize it, the better it will be for our environment. China is the largest buyer and producer of solar panels, as of 2015. The vast majority of solar panels are being installed in more remote areas in the form of giant solar farms. These farms sell energy to utility companies.

These giant farms that are popping up all over the country can be seen through many satellite images. The transition to solar power stems from the nation's high need for electricity due to the population size and the resultant air pollution. China's government is actively and aggressively encouraging financial institutions to provide incentives for solar installation.

Japan

Japan is highly dense with its population, based on land area. It does not have the luxury of covering huge geographic areas with solar panels. Solar farms, like those in China, are almost impossible to create. Still, Japan is among the leaders of the world in total solar energy production. In 2014, it produced about 23.3 gigawatts of the 177 gigawatts production worldwide.

Japan made a significant commitment to solar energy after the Fukushima nuclear plant disaster in 2011. Its plan is to double its renewable energy source by 2030. The country had to find creative places to put their solar panels because of the limited space it has. Since golf became unexpectedly popular in Japan during the 1980s, they decided to build an excessive amount of courses. Most of these were abandoned by 2015. These forgotten courses are now filled with numerous solar panels providing sufficient amounts of energy for areas around the country.

Japan became really inventive when it created solar islands with thousands of water-resistant solar panels. The great thing about these next-generation solar farms is their ability to be cooled more efficiently by the surrounding water.

Italy

Italy created about 18.5 gigawatts as of 2014, of the world's 177-gigawatt usage. This accounted for about 10% of the country's energy usage for that year. Solar energy usage has been expected to decline because the tax breaks for solar farms has expired, but Italy still stands in the top five for now.

United States

Billions of dollars in investments have been made by the United States in expanding its solar power dependence. A substantial increase in government incentives has also occurred in the residential sector, which is the fastest-growing segment of the market. Many homes are switching over to solar power after realizing the benefits this energy source has. In 2014, the US reported about 18.3 gigawatts of solar energy usage. This number is expected to rise considerably with the advent of new technology and prices that are competitive with nonrenewable resources, like fossil fuels.

Other countries that are major players in the solar power division include France, Spain, Australia, Belgium, and South Korea.

Best Solar Power in the United States

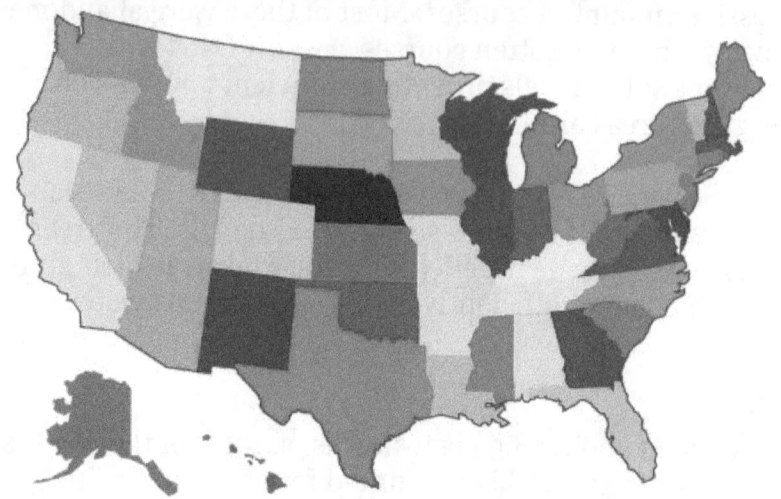

There are specific states in the US who are utilizing solar power at optimum levels. This is the result of geographical location and harnessing the power of the sun properly with the right resources. What states are doing solar the best as of 2019? Let's find out.

California

There is probably no surprise here. The golden state takes home the gold with solar power. The state gets a lot of sun year-round and is known to be very environmentally friendly. With 13,241 megawatts of solar capacity, it is capable of powering an estimated 3.32 million homes. California has more solar jobs and more megawatts of solar capacity in a single year than any other state.

Arizona

Once again, no surprise that Arizona comes in second place. The state boasts 2,303 megawatts of solar capacity, which is quite impressive. This is enough to power about 327,000 homes. Phoenix alone comes in at number three on the list of cities with the most installed solar photovoltaic capacity in the US. Even though the Arizona Corporation Commission has tried to restrict the distribution of solar energy, the state still maintains its place at number two.

North Carolina

North Carolina produces a lot of solar energy, with 2,087 megawatts of capacity that is capable of powering roughly 223,000 homes. The state is also creating a lot of solar jobs. These jobs are expected to grow in the coming years.

New Jersey

In New Jersey, they have installed 1,632 megawatts of solar capacity, which is enough to power 257,000 homes. The Garden State is proof that you don't have to be the sunniest place to take advantage of solar energy.

Nevada

The Silver State has the most solar capacity per capita in the US. As a desert, Nevada definitely gets a lot of sun on a yearly basis. Of course, the four seasons also exist in much of the state. With a solar capacity of about 1,240 megawatts, it can power about 191,000 homes.

Massachusetts

Massachusetts, on the bay, has a solar capacity of about 1,020 megawatts, which can power approximately 163,000 homes.

New York

Solar jobs in New York grew by about 13.3% in 2015. With 683 megawatts of solar capacity, it has the ability to power about 108,000 homes.

Hawaii

It is surprising that Hawaii is not higher on this list, but the city of Honolulu has installed more solar panels per capita than any other city in the country. The state of Hawaii has a total capacity of 564 megawatts, which can power about 146,000 homes.

Colorado

In 2015, about $305 million was invested in solar projects in the state of Colorado. The investment led to an additional 144 megawatts of solar capacity, bringing the total to 540 megawatts. That's enough solar power to affect 103,000 homes. Surprisingly, Colorado gets a lot more sun exposure than most people realize.

Texas

At number 10, Texas has a 534 megawatts capacity, giving it the ability to power up 57,000 homes. San Antonio alone ranked in the top seven for most solar-friendly cities. Texas gets a lot of sun exposure and has an abundance of open space. The potential for solar farms is huge in this state.

The numbers provided in this final US section were from the year 2015. If you live in one of these areas and want to switch to solar, then you are definitely not alone in your quest. The more solar capacity an area has, the easier it can be to get through multiple regulations. Whatever the case, I encourage you to find out the rules and regulations of your particular location and determine how you can get started with the transition to solar energy. Good Luck!

Conclusion

Thank you for making it through to the end of *Solar Power for Beginners;* let's hope it was informative and able to provide you with all of the tools you need to achieve your goals, whatever they may be. If you enjoyed it, I would be really curious to hear your opinion through a review, it means a lot to me!
The world has been reliant on fossil fuels for many decades as they have provided a quick and reliable source of energy in every aspect. Our lives are run by fossil fuels, from the homes we live in to the cars we drive and even the food we cook. Oil, gas, and coal are still the top three sources of energy around the world.

Unfortunately, these fossil fuels are harmful to the environment and can pose many health and safety concerns when not handled properly. There are alternatives to fossils fuels, like nuclear power, hydroelectric energy, or geothermal energy, but these come with their own issues, as well. Surprisingly, the energy source that is the safest and has the potential to provide us with more energy than ever though possible, is also the least utilized. I am referring to solar power.

The sun has been a fascinating topic for centuries. People slowly began understanding its viability and usefulness way before the common era. Solar energy has a rich history that is coming to light with more knowledge and information. With advanced technology, including solar panels and other high-level equipment, people have been able to harness the power of the sun and create a source of energy that is both safe, cheap, and renewable. When using solar power, our energy bills will be reduced significantly, and we will reduce our carbon footprint, effectively helping the environment.

The goal of this book, *Solar Power for Beginners,* was to describe in great detail what solar power is while comparing it to other energy sources and going over the massive benefits that it has. I also touched on some of the setbacks to make sure you are fully informed. After thoroughly understanding what solar power is, it was time to go over how to incorporate it into our own lives. The solar power equipment can be complex, but it can help us harness the energy from the sun and turn it into usable energy to use in our everyday lives.

I went over all of the equipment that is involved in a solar power system and tried to give you several options when I could. I also went over the step-by-step installation process to make sure you safely and securely install your equipment. I never want you to get harmed or mess up your equipment. I always advise that you hire a professional to help you so that you can have tranquility in knowing the panels were installed the right way. Of course, always do your research whenever you are hiring somebody.

I also provided information on companies that are leaders in the solar industry to give you an idea of where to get started and what to look for. You are certainly welcome to do your own research, as well. In fact, I encourage this.

Finally, I discussed some of the areas in the world that are doing solar the right way. The United States is certainly on this list and continuing to progress to solar power as the nation's chief energy source. My hope is that you developed a new understanding of solar power, and any doubts that you had were answered and addressed.
Despite continuous growth, solar power is still the most underutilized source of energy, and I believe that I can help change chant overtime. If we all work together, we can create a solar world that reduces our energy bill and helps save the world. It will take a lot of time and effort. We have collectively been dependent on fossil fuels and other nonrenewable energy sources for so long that it will take a

while to transition fully to renewable sources, like solar power.

The next step is to take the information provided in this book and start doing your own research. If you like what you've read about solar power so far, then don't stop here because there is still a wealth of information to obtain. Start studying up about what solar can offer you and set up a budget for a solar power system. Take all of the steps to ensure the process is as smooth as possible. If you are ready to take the step into renewable energy, then do not wait any longer to get started. Shift over to solar and begin lowering and even eliminating your energy bill while also reducing your carbon footprint.

Finally, a positive Amazon review is highly appreciated if you found this book helpful in any way! I want people around the world to benefit from this information, and a positive review on Amazon will go a long way in helping people become aware of it. Help me spread the word about the amazing benefits of solar power and renewable energy.

www.ingramcontent.com/pod-product-compliance
Lightning Source LLC
Chambersburg PA
CBHW070658220526
45466CB00001B/496